COMMUNICATION NETS

stochastic message flow and delay

COMMUNICATION NETS

stochastic message flow and delay

LEONARD KLEINROCK

Professor of Computer Science
University of California at Los Angeles

Formerly with Lincoln Laboratory
Massachusetts Institute of Technology

DOVER PUBLICATIONS, INC.
NEW YORK

Published in Canada by General Publishing Company, Ltd., 30 Lesmill Road, Don Mills, Toronto, Ontario.
Published in the United Kingdom by Constable and Company, Ltd., 10 Orange Street, London WC 2.

This Dover edition, first published in 1972, is an unabridged republication of the work originally published by McGraw-Hill Book Company in 1964.

International Standard Book Number: 0-486-61105-1
Library of Congress Catalog Card Number: 72-86227

Manufactured in the United States of America
Dover Publications, Inc.
180 Varick Street
New York, N.Y. 10014

to my parents,
Anne and Bernard Kleinrock

preface

The purpose of this monograph is to consider the stochastic flow of message traffic in connected networks of communication centers. The networks considered are channel-capacity limited, and consequently the measure of performance is taken to be the average delay encountered by a message in passing through the net. Questions pertaining to the assignment of channel capacities, effect of priority discipline, choice of routing procedure, and design of topological structure are considered.

For some years, there has been considerable research effort devoted to the formulation and solution of problems concerned with the flow of goods, people, water, etc., in connected networks. The emphasis has been on the study of constant deterministic flow through these nets. Quite aside from this, a large body of knowledge has been accumulating on the subject of queues; the emphasis there has been on stochastic flow through single-node servicing facilities. In our investigation, we combine the connected networks of the first study with the stochastic flow of the second study, emphasizing the operational rather than the combinatorial aspects of the system. Furthermore, owing to the intense interest in communication theory (which has come a long way in developing a general theory of message transmission between two communication centers), we direct our attention to connected networks of communication centers. Our intent is to provide a basis for understanding the general behavior and operation of such communication nets.

In Chapter 1 we discuss the large class of flow and transportation networks and distill from that class the particular form of communication net which is to be considered. In Chapter 2 we summarize the results obtained. The extreme complexity involved in obtaining exact solutions to the problem is demonstrated in Chapter 3, and an assumption necessary to the analysis is introduced and justified. The optimum channel capacity assignment for nets is presented in Chapter 4. Certain questions of queue discipline are considered in Chapter 5, and Chapter 6 deals with a class of random routing procedures. Finally, many of the separate facets of the problem are brought into focus simultaneously in Chapter 7, where the simulation of communication nets is discussed. In the concluding chapter we outline a number of extensions of the work which are worthy of investigation. An extensive set of appendixes is included which contains the details of proof for the various theorems presented in the main text.

This monograph is based upon a doctoral thesis submitted to the Department of Electrical Engineering of M.I.T. in December, 1962. Since the material is presented from first principles, the monograph is self-contained; Appendix A has been included for the reader unfamiliar with queueing theory. It is assumed that the reader is conversant with calculus and elementary probability theory.

I am happy to have this opportunity to express my sincere appreciation to Prof. Edward Arthurs for his encouragement and guidance in the supervision of the thesis upon which this monograph is based. I am also indebted to Dr. Herbert P. Galliher, Jr., whose work in the M.I.T. Operations Research Center was supported in part by the Army Research Office (Durham), and to Professor Claude E. Shannon, Donner Professor of Science at M.I.T., for their suggestions and ideas, which proved invaluable. In addition, I wish to express my thanks to the M.I.T. Lincoln Laboratory[1] for the interest and encouragement offered to me as a Staff Associate and for the use of the TX-2 digital computer. The informative discussions with numerous members of the M.I.T. and Lincoln Laboratory communities were extremely useful and are deeply appreciated.

Leonard Kleinrock

[1] Operated with support from the U.S. Air Force.

contents

appendixes

chapter 1

Introduction

We regularly encounter situations involving flow in our daily living. Consider, for example, the movement of automobile traffic, the transfer of goods, the streaming of water, the transmission of telephone or telegraph messages, and the passage of customers through the checkout counter of a supermarket. Although these examples represent rather distinct functional systems, they possess a fundamental property in common: they represent processes involving the flow of some commodity through a channel or a network of channels.

1.1 Systems of Flow: Examples

Let us begin by discussing some examples which illustrate the significant properties of various systems of flow. The simplest structural system consists of a single channel. The flow through that channel is made up of a pattern of arrivals (of some commodity), each of which comes up to the channel and requires or demands the use of that channel for a certain interval of time. It becomes immediately obvious that if we are to satisfy the demands placed on the system, we must ensure that the capacity of our channel is sufficient to handle the average rate of flow. If we have a *steady flow*,[1] then the problem is straightforward. For example, consider a department-store escalator

---

[1] By this we mean that the arrivals occur at uniformly spaced intervals of time and that the demands on the channel are the same for each arrival.

which can accept a steady input of up to C passengers per second, as shown in Fig. 1.1. If the rate R at which customers arrive at the escalator is constant (i.e., a steady flow), then either $R \leq C$, in which case no waiting lines build up and the steady flow is sustained, or else $R > C$, in which case the waiting line grows without bound and the system overflows.

Thus, steady flow through a single-channel system is easily disposed of. On the other hand, the case of *fluctuating* or *unsteady flow* through a single channel is a problem of considerable complexity. We illustrate this unsteady flow with a second example. Consider the familiar situation of a bakery store with one clerk at which customers arrive at unpredictable times. Before a customer is served, it is generally not clear exactly how long he will engage the clerk, and so his demand (or service time) on the channel (the clerk) is unpredictable. Owing to this uncertain or unsteady flow, we may expect a waiting line of customers to form in the bakery. We may assume, as is usually the case, that each customer receives a numbered ticket upon entering the store, so that a first-come first-served priority is maintained. Now, even if the clerk is able to keep up with the average demand of the customers, a waiting line will often form. Why this is so may easily be appreciated, for two customers entering the empty store, one directly following the other, immediately give rise to a waiting line of one customer. Thus, waiting lines form, in general, when a higher than

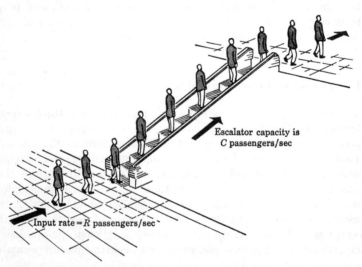

Escalator capacity is
C passengers/sec

Input rate $= R$ passengers/sec

Fig. 1.1. *Department-store escalator problem—an example of steady flow through a single channel.*

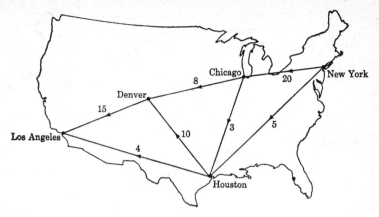

Fig. 1.2. *Diaper-transport problem, showing channel capacity.*

average flow occurs and overloads the clerk; on the other hand, when the flow is smaller than average, the clerk may find herself idle for some period. In any form of flow, whenever the average input rate exceeds the channel capacity, the waiting line grows without bound.

We now move ahead to the more involved case of a *network* of channels. Once again, we begin by considering steady flow. A simple example of such a net is shown in Fig. 1.2 and may be referred to as the *diaper-transport* problem. The diaper manufacturer in New York is faced with the problem of establishing a steady flow of newly produced diapers to his outlet in Los Angeles. The diapers may travel only along the routes shown in the figure. The number on each path indicates the maximum diaper tonnage that may pass along that path each week. Assuming that Los Angeles is in constant dire need of as many diapers as can be shipped, we wish to find the transport schedule which maximizes the steady flow from New York to Los Angeles. Problems of this general type have been carefully analyzed, and computationally efficient solution procedures exist [1].[1] Our immediate purpose in presenting this example is to convey the notion of a steady flow in a connected network of channels.

Let us now take the final step in our development and consider an example of unsteady flow in a connected network. We choose as an example a communication net whose topology (structure) is identical with the diaper-transport problem, but whose operation is significantly different. The function of this net is to transmit telegraph messages between the cities shown in Fig. 1.3.

---

[1] Numerals in brackets refer to the bibliography.

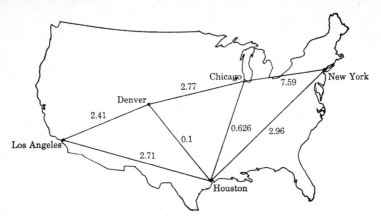

Fig. 1.3. *United States telegraph-network example.*

For graphical simplicity, each branch in this example represents a two-way connection (or channel) between cities, each with a capacity in binary digits (bits) per second equal to the number shown in the figure.   We recognize that the times of arrival of new telegrams at the various cities are, in general, unpredictable, as is the length of any particular telegram; consequently, we expect a waiting line of messages to form at each of the cities.   This is not unlike the situation in our bakery-store example.   Moreover, we are not certain as to where the next telegram will originate or what its destination will be.   Finally, we are faced with the additional complication that telegrams in transit may pass through intermediate relay cities, where they must be received, stored, and then retransmitted.

A man sending a telegram from New York to Los Angeles may wish to know how long the transmission is expected to take.   A large national airline may be interested in installing a private telegraph communication net; they would like to know how much capacity will be required on each channel so as to provide an acceptable average delay for a given traffic density.   We propose to answer these and other questions in succeeding chapters.

1.2   A Closer Look

Up to this point, we have given heuristic descriptions of some flow problems and have found that important differences exist among them. As a result, it is necessary for our purposes to classify all systems of flow into four distinct groups: steady flow through a single channel, unsteady flow through a single channel, steady flow through a network of channels, and unsteady flow through a network of channels.

Let us return to our four examples again and examine more carefully the class of problems from which each is drawn. The department-store escalator example is representative of single-channel systems subject to steady flow inputs. The solution of these systems is, as we have seen, trivial, and in fact they represent a special case of the more general class of single-channel systems subject to unsteady flow. Therefore, we pass immediately to this second class of single-channel systems, which we have illustrated with our bakery-store example. As indicated, this problem is hardly trivial. Indeed, a substantial theory of waiting-line (or queueing) processes exists which deals with the description and analysis of the effects of unsteady (or stochastic) flow. By *stochastic flow* we mean specifically that both the time between successive arrivals to the system and the demand placed on the channel by each of these arrivals are random quantities. We recognize that the special case of a stochastic flow in which the inter-arrival times are all equal and in which the service demands of each customer are the same reduces to our original steady-flow problem. That is, steady flow is a special case of stochastic flow. However, as opposed to steady flow, a queue may form during stochastic flow even if the average flow rate does not exceed the capacity of the channel; in such a case, the average queue length remains finite. In Appendix A we derive some of the results of queueing theory which we shall use in our analysis of communication nets. We suggest a careful reading of this appendix to the reader unfamiliar with queueing theory.

One of the earliest investigations of a queueing process dates back to 1907, when E. Johannsen[1] published two papers concerning delays and busy signals in telephone traffic problems. Under his influence, A. K. Erlang[2] undertook the consideration of similar problems, and an English translation and summary of Erlang's major contributions may be found in [2]. A number of others also contributed to the early development of queueing theory [3,4,5,6]. In 1928, T. C. Fry [7] published his book (which has since become a classic work) surveying and unifying the approach to congestion problems up to that time. In 1939, Feller [8] made an outstanding contribution to queueing theory through his presentation of the birth and death process.

In the early 1950s, it became obvious that many of the results obtained in the field of telephony were applicable to more general

---

[1] Reference to Johannsen's work will be found in [2].

[2] Erlang was an engineer with the Copenhagen Telephone Exchange.

situations; thus started investigations into waiting lines of many kinds which have developed into modern queueing theory (a theory which finds numerous applications in the field of operations research).   The emphasis has been on single-channel facilities, i.e., systems in which "customers" enter, join a queue, eventually obtain "service," and upon completion of this service leave the system, as in our bakery-store example.   D. G. Kendall published two notable papers [9, 1951, and 10, 1953] in which he made substantial contributions to the theory. In 1952, D. V. Lindley [11] derived an integral equation describing queueing processes with arbitrary arrival and service distributions. A number of other researchers have made contributions to the modern theory of queueing [12,13,14,15,16,17].   Many books have since been published which summarize much of the current work in queueing, offering both applications and certain analytic aspects of the theory. Among these books are W. Feller [18, 1950], F. Pollaczek [19, 1957], P. M. Morse [20, 1958], A. Ja. Khinchine [21, 1960], R. Syski [22, 1960], T. L. Saaty [23, 1961], D. R. Cox and W. L. Smith [24, 1961], and J. Riordan [25, 1962].   This vast body of knowledge concerning queues allows us to gain a rather thorough understanding of single-channel flows.   We draw heavily from this knowledge in our study of communication nets.

We now return to the diaper-transport problem as an example of steady flow in a connected network of channels.   The problem as set forth earlier is representative of a large class of problems referred to as *static maximal flow* problems.   Ford and Fulkerson [1] describe solutions to a large number of problems that have formulations in terms of steady flow through capacity-limited networks.   Most of the problems they discuss fall into two categories.   The first, static maximal flow problems, is concerned with determining the maximal steady flow between two points in a network whose channel capacities have already been chosen.   The second, *minimal cost flow* problems, deals with the method by which one designs the network flows so as to minimize cost while satisfying certain flow requirements of some of the nodes.   The solution to the static maximal flow problem is embodied in a fundamental result known as the *max-flow min-cut theorem*.   The theorem states that the maximum steady flow between any two nodes in a net is equal to the minimum cut capacity of all cuts which separate the two nodes.   A *cut* is defined as a set of branches which, if removed from the net, would leave two unconnected nets; the *cut capacity* is defined as the sum of the capacities of all branches which have been removed by the cut.   As an example, one cut which separates the net in Fig. 1.2 into two pieces (one containing New York, and the other

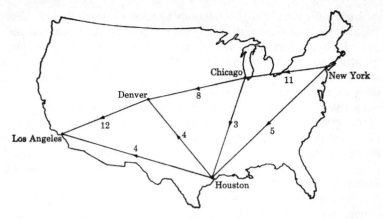

Fig. 1.4. *Diaper-transport problem, showing a maximal-flow solution.*

containing Los Angeles) consists of the branches between Chicago and Denver, Houston and Denver, and Houston and Los Angeles; the capacity of this cut is easily seen to be 22 tons of diapers per week. For a proof of this theorem we refer the reader either to Ford and Fulkerson [1] or Elias, Feinstein, and Shannon [26]. Any network flow which achieves this calculable maximal flow is therefore a solution. For the sake of completeness, we present in Fig. 1.4 a solution to the diaper-transport problem which achieves the maximal flow of 16 tons of diapers per week.[1] Solutions to the minimal cost flow problem (also referred to as the *transportation* or *Hitchcock* problem) are presented in [1].

An important class of problems which concern themselves with satisfying a multiplicity of flow requirements has also been investigated. Such a problem is referred to as a *multiterminal network flow* problem if the pairwise requirements are to be satisfied individually. If, on the other hand, the pairwise requirements are to be satisfied simultaneously, it is referred to as a *multicommodity network flow* problem. The former is carefully discussed in the book by Ford and Fulkerson [1], in which are described the contributions of a number of people, notably Gomory and Hu [27,28], Chien [29], and Mayeda [30]. Solutions to the multicommodity problem are in an earlier stage of development; contributions to date have come from Gomory and Hu [31], Jewell [32], Robacker [33], and Hakimi [34]. Unfortunately,

---

[1] The number above each channel represents the flow rate for that channel.

it is generally true that the solutions to these network flow problems take the form of algorithms with vast computational requirements.

In 1956, Z. Prihar [35] presented an article in which he explored certain topological properties of communication nets; for example, he showed matrix methods for finding the number of ways to travel between two nodes in a specific number of steps.

Lastly, we come to the example of stochastic flow through a telegraph network in the United States. In contrast to the abundant supply of papers both in queueing theory (on stochastic flow through single channels) and in network flow theory (on steady flow through connected nets), relatively few works have been published on connected nets subject to stochastic flow (which is the area of interest of this monograph). Among those papers which have been presented is one by G. C. Hunt [36] in which he considers sequential arrays of waiting lines. He presents a table which gives the maximum utilization factor (ratio of average arrival rate to maximum service rate) for which steady-state probabilities of queue length exist, under various allowable queue lengths between sequential service facilities. J. R. Jackson [37] published a paper in 1957 in which he investigated networks of waiting lines. His networks consisted of a number of service facilities into which customers entered both from external sources as well as after having completed service in another facility. He proved a theorem which, stated roughly, says that a steady-state distribution for the system state exists as long as the effective utilization factor for each facility is less than 1, and in fact, this distribution takes on a form similar to the solution for the single-node case. In 1950, C. E. Shannon [38] considered the problem of storage requirements in telephone exchanges and concluded that a bound can be placed on the size of such storage by estimating the amount of information used in making the required connections. In 1959, P. A. P. Moran [39] wrote a monograph on the theory of storage. The book describes the basic probability problems that arise in the theory of storage, paying particular attention to problems of inventory, queueing, and dam storage. It is one of the few works pertaining to a system of storage facilities. In 1962, R. Prosser [40] offered an approximate analysis of a random routing procedure in a stochastically fed communication net in which he shows that such procedures are highly inefficient but extremely stable (i.e., they degrade gracefully under partial failure of the network). Furthermore, Prosser [41] describes an approximate analysis of certain directory procedures in which he concludes that the disadvantages of such procedures are the necessity of maintaining the directories and the need to determine optimal paths from the directory information; he also concludes that the advantages (as compared to random routing procedures) are the increase in efficiency and in the capacity of operation.

1.3 The Emphasis

By means of our four graphic examples, we have considered an extensive class of flow systems. Beginning with the simple case of steady flow through a single channel, we have carried the discussion to extremely complex systems of stochastic flow through networks of channels. It is clear from Sec. 1.2 that considerable research effort has gone into the investigation of many of these flow systems, viz., queueing theory, linear programming, and network flow theory. On the other hand, there is a general paucity of knowledge concerning networks subject to stochastic inputs. The reason for this contrast is clear: the latter problem is extremely difficult. It is this problem to which we address ourselves.

Specifically, this monograph focuses attention on the stochastic flow of message traffic through connected networks of communication centers. We direct our remarks to networks of communication centers (as opposed to other network applications) in view of the fact that communications and communication theory[1] have come to play such a crucial role in modern engineering systems.

The fundamental characteristic of these communication nets is that queues of messages form at each of the network's nodes, owing to the sporadic or stochastic nature of the message flow. Consequently, each node must provide some form of storage for the messages as they wait on the queue.

The single most significant measure of performance for these nets is the average time for a message to make its way from its origin through the net to its destination. We refer to this as the *average message delay*. The remainder of this monograph is essentially devoted to the optimization (i.e., minimization) of this performance measure for a variety of communication nets.

---

[1] The results from information theory [42] have a strong relation to the communication net problems considered here. Most of the work in information theory has dealt with communication between two points, rather than communication within a network. In particular, one of the major results says that there is a trade-off between message constraint length and probability of error in the transmitted message for transmission over noisy channels. The effect of this constraint length is to add additional intranode delays to the message. Thus, if delays are of no consequence, transmission with an arbitrarily low probability of error can be achieved. We do not deal specifically with noisy channels, although such an avenue of investigation represents an interesting extension for future study.

1.4 Elementary Concepts

For the introduction of the elementary concepts associated with our communication nets, it is useful to carry along our example of the United States telegraph network shown in Fig. 1.3. In this figure, the *nodes* represent *communication centers*, which, in our example, correspond to switching centers in cities of the United States; they could, however, correspond to switching centers in, say, communication satellites in space. The ordered[1] connections, or *links*, between the nodes represent one-way *communication channels*, each with its own channel capacity. For our purposes, *messages*, which must pass through the net, consist of the specification of the following quantities: the node of origination, the destination node, the time of arrival to the network, the message length in bits,[2] and the message priority class. In general, these stochastic quantities are specified according to some probability distribution. As an example, suppose that a message originates at Los Angeles in Fig. 1.3 at time $t = 0$ and has for its destination New York; let its length be 271 bits, and assume that we have no priority structure associated with the messages. Let us follow this test message across the United States through the network. At the switching center in Los Angeles, a decision must be made as to whether to send the message to Denver or to Houston. The decision rule is referred to as a *routing procedure* and is, in general, a function of the current state of the net. Channels leaving and entering a node may be used independently and simultaneously, each one for a different message. Thus, when the test message enters Los Angeles, it may find zero, one, or two channels in the process of transmitting other messages. If all channels are busy, then the message joins a queue; this is accomplished physically by storage of the message in a memory. The notion of queues of messages forming at the nodes is basic to the communication nets under consideration; we may thus think of the communication net as a network of queues. When the message reaches the front of the queue, the routing procedure is used to decide

---

[1] As mentioned earlier, the figure shows unordered connections (i.e., connections with no directional arrows); these are to be interpreted as two-way links, each independent of the other and each with a capacity equal to the number shown adjacent to the link. That is, for graphical simplicity we choose to represent the fully duplexed channels ⟳ with the symbol •——•

[2] In transmitting messages, we are concerned with the data rate of transmission, which is not necessarily the information rate in the information-theoretic sense.

which channel the message will be sent over. Let us assume that the channel connecting Los Angeles to Houston is chosen. Since the capacity of this channel is 2.71 bits/sec, our message will spend 100 sec in transmission. Clearly, no other message may use the channel during this time. When the transmission is completed, this channel may accept a new message from the queue for transmission. Upon reception in Houston, the process which took place at Los Angeles is essentially repeated, and the message may have to wait in a queue (if all or some of the channels leaving Houston are busy). Eventually, however, the message will make its way through the net to its destination in New York. When it arrives there, it is considered to be dropped from the net.

It is now clear why these nets are often referred to as *store-and-forward* communication nets; viz., in passing through a node, the messages are stored, if necessary, and then forwarded (transmitted) to the next node on the way to their destination. The total time that a message spends in the network is referred to as the *message delay*. We now introduce the concept of a *traffic matrix* whose ij entry describes the average number of messages generated per second which have node i as origin and node j as destination. The priority classes referred to previously merely dictate the way in which the messages in a queue are ordered (clearly, preferential treatment is given to higher-priority messages).

In summary, then, we have introduced the following:

1. *Node:* a communication center which receives, stores, and transmits messages
2. *Link:* a one-way communication channel
3. *Network:* a finite collection of nodes connected to each other by links
4. *Message:* specified by its origin, destination, origination time, length, and priority class
5. *Routing procedure:* a decision rule which is exercised when a message is routed from one node to another
6. *Queue:* a waiting line (composed of messages in our case)
7. *Queue discipline:* a priority rule which determines a message's relative position in the queue
8. *Message delay:* the total time that a message spends in the net
9. *Traffic matrix:* the ij entry in this matrix describes the average number of messages generated per second which have node i as an origin and node j as a destination

Having introduced the elementary concepts, we now inquire into those quantities which are of interest in our study of communication nets. We consider these quantities from the viewpoint of the user, the operator, and the designer of the net. Specifically, the user (i.e., the originator and recipient of messages) is concerned with:

1. The average message delay
2. The total traffic-handling capability of the net

The operator (i.e., the one who controls the flow of messages through a node) is concerned with:

1. The routing procedure
2. The priority discipline
3. The storage capacity at each node

The designer of the net is interested in:

1. The average message delay
2. The total traffic-handling capability of the net
3. The routing procedure
4. The priority discipline
5. The storage capacity at each node
6. The channel capacity of each link
7. The topological structure of the net
8. The total cost of the system

As expected, the designer's interest includes and extends beyond those quantities of interest to the user and operator. We choose, therefore, to investigate all these quantities, as well as certain trading relations which exist among some of them.

1.5 An Existing Store-and-Forward Communication Net

In this section we describe an existing store-and-forward communication net. In Sec. 1.6 we abstract a mathematical model to represent systems of this type for purposes of analysis. We choose for this description an automatic telegraph switching system (Plan 55A [43]) which has been developed by Western Union[1] for the Air Force to handle large quantities of military traffic over a worldwide network. The system was recently installed and consists of 10 switching centers (five domestic and five overseas). These switching centers are interconnected through a network of lines or radio channels which comprise the communicating system for automatic relay of telegraph messages. In addition, each of the main switching centers is con-

---

[1] This material has been condensed from the *Western Union Technical Review* with the express permission of the Western Union Telegraph Company.

nected by lines to a set of tributary stations in the region served by that center. Messages originate at the tributary stations and are transmitted to the regional switching center, and then perhaps to other switching centers, where, finally, they are transmitted to their destinations at other tributary stations.

In the switching centers of this system (i.e., Western Union's Plan 55A), messages are received and retransmitted in the form of punched (perforated) paper tape. A message's destination is controlled by routing indicators (normally groups of six letters) recorded on the paper tape as part of the message heading. The switching of messages takes place automatically, except at the points of origin and destination. However, it is possible to convert to manual (push-button) switching at each center at any time; this mode of operation is abnormal and is used only in case of failure in the automatic switching devices or in cases of improper format in the received messages.

A strict priority or precedence structure is included in the system, and messages are transmitted in this order of precedence. Six priorities are used in the system and are detected by inspection of two letters, referred to as *precedence prosigns*, in the message heading.

In each switching center through which it passes, a message is perforated and transmitted twice. The first reperforation takes place as the message is being received into the center. The message is then switched and transmitted across the office to a transmitting (or sending) line position, where it is reperforated and transmitted again. The perforated paper tape serves as the store or buffer within the switching center. In cross-office transmission, the cross-office line connections are not established until the entire message has been received into the center (i.e., until the end-of-message symbols are received); the only exception to this rule comes about when emergency messages are received, in which case the connections are set up immediately. From the sending position, the message is either sent to a tributary (and therefore its destination) or to another switching center. The presence of more than one routing indicator in a message heading indicates a multiple-address message. A message of this type is processed in such a way that an individual copy reaches each destination.

The incoming cabinet and the outgoing cabinet are the two principal pieces of equipment in a switching center. These cabinets are linked together by cross-office channels (switching circuits) which carry signals at a rate of 200 words per minute (wpm). Electronic pulses on a single conductor are used as cross-office transmission signals (as opposed to the older torn-tape system which required an attendant to tear the tape off the receiving apparatus, carry this tape across the office, and insert the tape into an appropriate transmitter). For those intercenter channels which carry heavy traffic loads, several interconnecting line circuits and sending positions are sometimes required. All sending positions in such a multiple-circuit group transmit to identical destinations. Any message for such a destination can be switched to any idle circuit within the group.

Messages are received into the center at 60 or 100 wpm; they are then transmitted cross-office at 200 wpm and, finally, are retransmitted to outgoing lines at 60 or 100 wpm. Thus, since the cross-office rate is at least twice that of the outgoing lines, a sufficient number of messages can be sent across the office to keep the outgoing lines busy most of the time, and the cross-office transmitters are idle at least half the time. This being the case, the receiving positions seldom find it necessary to wait for a cross-office connection, and no large quantity of backlogged paper tape should form at these positions. If a backlog develops, the higher cross-office rate should quickly relieve the situation once a cross-office connection is obtained.

When a receiving position has a message that is to be switched to an outgoing circuit which is busy, the message must wait until a circuit to the desired destination becomes available; if the wait is excessive, or if the message is of extremely high priority, an alarm is operated which calls an attendant to the position to take suitable action.

This completes our description of one existing store-and-forward communication net. Although it does not include in its description all current procedures or equipment, this system does exemplify many message-switching nets.

1.6 The Model; Assumptions

The description in Sec. 1.5 provides us with an existing store-and-forward communication net from which we may abstract a meaningful idealized mathematical model. The motivation for using an idealized model is simply that of mathematical ease and tractability; at the same time, however, we must ensure that the idealizations introduced lead to a model which retains the essential characteristics of the real system. Recall that we have chosen the *average message delay* as our measure of network performance. Accordingly, we desire that our model, although idealized, exposes the fundamental behavior of the average message delay in store-and-forward nets.

Consider the elementary concepts presented in Sec. 1.4. We offer that description as a starting point for our model, and we now proceed to apply certain assumptions to it. Specifically, the nodes refer to the switching centers, and we consider that the tributary (or originating) stations are part of the switching center itself. We assume that all channels are noiseless and that all communication centers (nodes) and channels are invulnerable to damage or destruction (the reliability question). This assumption implies that there are no theoretical or practical problems in transmitting over the channel at a data rate equal to the channel capacity. That is, we may assume that the messages have been encoded into a binary alphabet so that each binary digit corresponds to one bit of data to be transmitted. The encoding required to reduce errors in a noisy channel would introduce additional intranode delays to the message; we do not consider this case. Furthermore, we assume that cross-office delays are negligible as compared to the channel transmission time (a reasonable assumption based upon information on existing and proposed systems).

The study is restricted to data or message traffic as distinct from telephone or direct-wire traffic, which has not been considered. We assume that each message has a single destination (as opposed to an all-points message, for example) and that each message must reach that destination before leaving the network (i.e., no defections); this involves the additional assumption of an unlimited storage capacity at each node to supply a "waiting room" for the messages in the queue.

In transmission between two nodes, a message is considered to be received at the second node only after it is fully received. The consequence of this assumption is that messages may not be retransmitted out of a node while they are being received into the node. Clearly, this represents, at worst, a slightly conservative assumption as regards the message delay in a node. Moreover, many store-and-forward nets do indeed operate in this manner, because of the difference in capacity between incoming and cross-office channels.

The origination times and lengths of the population of messages which will flow through the net cannot be predicted beforehand with complete accuracy. We may, however, describe these random variables statistically by means of their probability density functions. Specifically, we assume that the origination times and lengths of messages are chosen individually and collectively at random; this, of course, implies that the interarrival times between messages are exponentially distributed (i.e., Poisson), as are the message lengths. Furthermore, we assume stationarity for these stochastic processes that feed the net.

We have introduced a number of assumptions in arriving at a model for a store-and-forward communication net. We summarize these as follows:

1. Noiseless channels and perfectly reliable nodes and channels
2. Negligible cross-office delays
3. Restriction to message traffic
4. Single destination for each message, with no defections
5. Infinite storage capacity at each node
6. Full reception of message before retransmission
7. Poisson arrival statistics
8. Exponentially distributed message lengths
9. Stationarity and independence of the stochastic processes 7 and 8

Whereas assumptions 1 to 6 do not correspond exactly to the true situation in any real net, they do describe idealizations which are reasonable and close to reality.

Assumptions 7 to 9 are of a somewhat different nature. In particular, they assign specific distributions to message arrival times and lengths. There is available no quantitative data from which to determine the actual form of these distributions. However, certain data obtained by Molina [3] for telephone traffic correspond very well to these assumptions. Moreover, these distributions avoid considerable mathematical complication and, at the same time, correspond to reasonable (and perhaps conservative) assumptions.

It is appropriate to mention here that some of the results presented in this work include the assumption of a *constant* total channel capacity assigned to the net (i.e., the sum of the capacities of all channels in the net is held fixed). Finally, we note that one additional assumption is required before we arrive at a mathematically tractable model; we delay discussion of this final assumption until Chap. 3.

In summary, we state again that the worth of this model lies mainly in its retention of the essential character of the message delay in real store-and-forward communication nets.

1.7 Notation and Further Definitions

As a matter of convenience, we define and list below some of the important quantities and symbols.

γ_{jk} = average number of messages entering network per second with origin j and destination k

λ_i = average number of messages entering ith channel per second

$1/\mu_{jk}$ = average length of messages which have origin j and destination k, bits

C_i = channel capacity of ith channel

γ = total arrival rate of messages from external sources (see below)

λ = total arrival rate of messages to channels within net (see below)

$\bar{n}$ = average path length for messages (see below)

$1/\mu$ = average message length from all sources (see below)

C = sum of all channel capacities in net (see below)

ρ = network load, namely, ratio of average arrival rate of bits into net from external sources to total capacity of net (see below)

Z_{jk} = average message delay for messages with origin j and destination k

T_i = average delay for a message passing through channel i (includes both time on queue and time in transmission)

T = average time messages spend in network (see below), taken as measure of performance of a net

τ = traffic matrix with entries γ_{jk}

We collect below certain relations among the definitions above. Some of these relations are defined, and others may be obtained by simple manipulation.

$$\bar{n} = \frac{\lambda}{\gamma} \qquad \text{[see Eqs. (B.12) and (B.13)]}$$

$$\frac{1}{\mu} \equiv \sum_{j,k} \frac{\gamma_{jk}}{\gamma} \frac{1}{\mu_{jk}}$$

$$C \equiv \sum_{i} C_i$$

$$\gamma \equiv \sum_{j,k} \gamma_{jk}$$

$$\rho \equiv \frac{\gamma}{\mu C}$$

$$\lambda \equiv \sum_{i} \lambda_i$$

$$T \equiv \sum_{j,k} \frac{\gamma_{jk}}{\gamma} Z_{jk} = \sum_{i} \frac{\lambda_i}{\gamma} T_i \qquad \text{[see Eqs. (4.16) and (4.17)]}$$

1.8 The Precise Problem Statement

At the outset (indeed, in the preface) we indicated that our interest in communication nets was directed toward questions pertaining to the assignment of channel capacity, effect of queue discipline, choice of routing procedure, and design of topological structure.   Subsequently, we defined the measure of performance to be the average message delay.

Once we focus our attention on a performance measure, we may no longer be content with the mere analysis of communication nets; rather, we must insist on an *optimal* solution to the network design problem.   We define an optimal solution to be one which minimizes the average message delay at a fixed network cost.   The fixed cost function D may be described by the following equation:

$$D = \sum_{i=1}^{N} d_i C_i \qquad (1.1)$$

where C_i is the channel capacity of the ith transmission channel, and d_i is a function (independent of the capacity C_i) which reflects the cost, say in dollars, of supplying one unit of channel capacity to the ith channel.   The quantity D represents the total number of dollars available to spend in supplying the N-channel system with the set of capacities C_i.   The actual values of the set d_i depend upon the particular communication net involved; for example, d_i might be chosen to represent the length of the ith channel.   This particular form of cost function associates the entire network cost with the channels themselves.   This represents no loss of generality as regards the nodal costs since, although more difficult to evaluate, they may be grouped with the channel costs; this grouping of costs results in a reduction of mathematical complexity.

We have thus described a set of design parameters which may be adjusted, a performance measure to be minimized, and a fixed cost constraint.   This defines our goal.   The problem statement may be expressed as

$$\text{Minimize} \atop {\left\{ \begin{array}{l} \text{Capacity assignment,} \\ \text{routing procedure,} \\ \text{priority discipline,} \\ \text{topology} \end{array} \right.}} \quad T = \sum_{i=1}^{N} \frac{\lambda_i}{\gamma} T_i \qquad (1.2)$$

subject to
$$D = \sum_{i=1}^{N} d_i C_i$$

where N is the number of channels in the net (N may vary as the topology varies).

The precise problem expressed above has not been solved in its entirety.   In order to make progress toward a solution, we have been forced to freeze certain variables in the system (such as the routing procedure in some cases).   In the next chapter, we present a summary of results and discuss the particular compromises we have been obliged to make.

chapter **2**

Summary of Results

2.1 Analytic Results

The model we have adopted for connected networks of communication centers is described in Secs. 1.4 and 1.6. This model leads to a rather complex mathematical structure. We have, therefore, found it necessary to modify the original model with the introduction of the *independence assumption*. This assumption is carefully discussed in Chap. 3; in essence, it specifies that new lengths are chosen (from an exponential distribution) for messages each time they enter a node. As shown in Chap. 3, the new model results in a mathematical description which accurately describes the behavior of the message delay in the original model. As a consequence of the independence assumption and of Theorem A.1 (due to Burke), we are able to analyze each node separately in calculating message delay. We then find ourselves in a position to make some positive statements regarding the quantities of interest which are described in Sec. 1.4.

The results obtained from this research for the model described above (including the independence assumption) will now be summarized. We begin by considering a single node within the net, and we obtain several new results for single-node systems as described in Chap. 4. Specifically, if one considers the problem of finding the number N of output channels from a single node which minimizes the time that a message spends in the node (queueing time plus transmission time), subject to the constraint that each channel is assigned a capacity equal to C/N, one then finds (Theorem 4.2) that $N = 1$ is the optimum solution. Further, a new interpretation of the utilization factor[1] for a single node with multiple output channels is obtained. The obvious trading relations between message delay, channel capacity, and total traffic handled are also developed.

At this point, a result is obtained which has great bearing on the general network problem. It gives the assignment of channel capacity to a net consisting of N independent nodes (each with a single output channel; see Fig. 4.4), which minimizes the message delay averaged over the set of N nodes, subject to the constraint that the sum of the assigned channel capacities is constant.[2] Specifically, if λ_i is the average (Poisson) arrival rate of messages to the ith node, and $1/\mu_i$ is the average length of these messages (exponentially distributed), then the optimum assignment C_i of the channel capacity to the ith node is

$$C_i = \frac{\lambda_i}{\mu_i} + \left(C - \sum_{j=1}^{N} \frac{\lambda_j}{\mu_j} \right) \frac{\sqrt{\lambda_i/\mu_i}}{\sum_{j=1}^{N} \sqrt{\lambda_j/\mu_j}} \qquad (2.1)$$

where C is the fixed total capacity. Using this optimum assignment, we find that[3]

$$T = \sum_{i=1}^{N} \frac{\lambda_i}{\lambda} T_i = \frac{\left(\sum_{i=1}^{N} \sqrt{\lambda_i/\lambda\mu_i} \right)^2}{C(1 - \rho)} \qquad (2.2)$$

Here, T_i is the average message delay in the ith node, and T is the message delay appropriately averaged over the index i; recall that $\rho = \gamma/\mu C$. Theorem 4.4 considers minimizing T [as expressed in Eq. (2.2) above] with respect to the λ_i (assuming $\mu_i = \mu$ for all i), holding λ constant, subject to the additional constraints that $\lambda_i \geq k_i$ (where we take $k_1 \geq k_2 \geq \ldots \geq k_N$ with no loss of generality). The set of numbers k_i represents lower bounds on the traffic flow through each channel and corresponds to one form of physical limitation that may exist. The distribution of λ_i which minimizes T is

$$\lambda_i = \begin{cases} \lambda - \sum_{j=2}^{N} k_j & i = 1 \\ k_i & i > 1 \end{cases} \qquad (2.3)$$

---

[1] The utilization factor is merely the ratio of average arrival rate of bits into the node to the maximum transmission rate of bits out of the node.

[2] Note that this may be interpreted as a special case of the cost function expressed in Eq. (1.1), viz., where $D = C$ and $d_i = 1$ for all i.

[3] Note that the double subscript jk in γ_{jk} may in this case (see Fig. 4.4) be replaced by a single subscript i; thus, according to Sec. 1.7, $\lambda_i = \gamma_i$ in this special case, and $\lambda = \gamma$.

For all $k_i = 0$, all traffic is assigned to (any) one of the channels, and, by Eq. (2.1), this channel is allotted the total capacity C. In any case, we observe that this solution displays an attempt to *concentrate* the traffic as much as possible. In fact, the results expressed by Theorem 4.2 and by the trading relations of Sec. 4.3 also indicate the desirability of concentrating the traffic (and therefore the channel capacity) in order to minimize message delay.

If we now consider the general case of an interconnected net (as, for example, in Fig. 1.3) with N channels indexed by the subscript i, subject to a fixed routing procedure,[1] then we find that Eq. (2.1) continues to describe the optimum channel capacity assignment. λ_i is still the average arrival rate of messages to the ith channel; however, for this case, we take $\mu_i = \mu$ for all i. We choose to rewrite this equation in the following simplified form:

$$C_i = \frac{\lambda_i}{\mu} + C(1 - \bar{n}\rho) \frac{\sqrt{\lambda_i}}{\sum_{j=1}^{N} \sqrt{\lambda_j}} \tag{2.4}$$

Here $\bar{n}$ is the average path length for messages in the net. Furthermore, the average message delay T, under this optimum assignment, becomes

$$T = \frac{\bar{n} \left(\sum_{i=1}^{N} \sqrt{\lambda_i/\lambda} \right)^2}{\mu C(1 - \bar{n}\rho)} \tag{2.5}$$

The full significance of these last two equations is discussed below in conjunction with the summary of the simulation experiments. However, their importance can be emphasized by considering our telegraph network example (see Fig. 1.3). For a numerical evaluation of the optimum conditions expressed in Eqs. (2.4) and (2.5), we must decide upon some traffic matrix τ. We would like to have realistic values for this matrix, but unfortunately no quantitative data are readily available for this purpose. Consequently, we choose to make use of a conjecture put forward by Zipf [44]. Essentially, he surmises that the magnitude of the flow γ_{jk} between two cities whose populations are P_j and P_k and which are separated by a distance D_{jk} may be expressed as follows:

$$\gamma_{jk} = \alpha \frac{P_j P_k}{D_{jk}} \tag{2.6}$$

[1] By *fixed routing procedure* we mean that, given a message's origin and destination, there exists a unique path through the net which this message must follow. If more than one path is allowed, we speak of an *alternate routing procedure*.

where α is a factor of proportionality. He supports this conjecture by demonstrating its applicability in a wide variety of situations (e.g., the flow of telephone calls, the shipment of goods via railway express, and the quantity of mail flow between various cities). For the purposes of this example we may accept Eq. (2.6). By referring to a table of airline distances [45], we may determine the D_{jk}. Similarly, we may refer to the 1963 edition of the World Almanac to find the population P_j of the various metropolitan areas in our examples. We observe that, since $D_{jk} = D_{kj}$, the flow in Eq. (2.6) must be symmetric in j and k (that is, $\gamma_{jk} = \gamma_{kj}$). In Eq. (2.7) we give the traffic matrix τ obtained from this data (with $\alpha = 10^{-10}$ for convenience). From this we may

$$\tau =$$

	New York	Chicago	Houston	Los Angeles	Denver
New York		9.34	0.935	2.94	0.610
Chicago	9.34		0.820	2.40	0.628
Houston	0.935	0.820		0.608	0.131
Los Angeles	2.94	2.40	0.608		0.753
Denver	0.610	0.628	0.131	0.753	

$$(2.7)$$

observe that $\gamma = 38.33$. For simplicity we also choose $C = 38.33$ and $\mu = 10$. As far as the fixed routing procedure is concerned, we adopt the most obvious one, which consists of the set of shortest routes (e.g., Los Angeles routes through Houston to get to New York); we shall not bore the reader with its description. As a consequence of this routing procedure, we find that $\bar{n} = \lambda/\gamma = 1.31$.

We are now ready to make an evaluation of the channel capacity assignment described in Eq. (2.4) (hereafter referred to as the *square root channel capacity assignment*). To do this properly, we must observe the behavior of the average message delay both with the square root channel capacity assignment and with some other *plausible* capacity assignment. An intuitively reasonable assignment is one which allots a fraction of the total capacity C to each channel in direct proportion to the traffic carried by that channel, viz.,

$$C_i = \frac{\lambda_i}{\lambda} C \qquad (2.8)$$

We refer to this as the *proportional channel capacity assignment.* Indeed, we find that this is a rather good choice; although we have

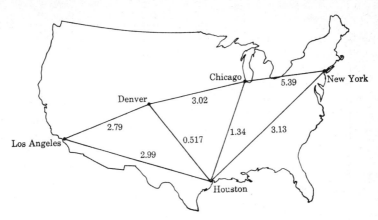

Fig. 2.1. *United States telegraph-network example with the optimum channel capacity assignment.*

shown that it is not optimum, it is nearly so. One observes that when all the λ_i are identical, the two channel capacity assignments give identical results; the same holds true when $\rho \rightarrow 1/\bar{n}$. Applying this proportional assignment to our example, we come up with the set of channel capacities which appear in Fig. 1.3. On the other hand, application of the square root channel capacity assignment results in the set of numbers shown in Fig. 2.1.

For the values of γ_{jk} in Eq. (2.7), we find that $\rho = 0.1$. In order to sweep ρ through its range, we need merely vary α. In so doing, we are finally able to calculate T as a function of ρ for both the square root and proportional channel capacity assignments. The results of these calculations are presented in Fig. 2.2. Observe that the square root assignment is superior to the proportional assignment (as of course it must be). It is easy to show that, for the proportional assignment,

$$T_{\text{prop}} = \bar{n}N/[\mu C(1 - \bar{n}\rho)];$$ the ratio T_{prop}/T is thus $N/\left(\sum_{i=1}^{N} \sqrt{\lambda_i/\lambda}\right)^2.$

This ratio is always greater than or equal to unity,[1] equality being obtained when all the λ_i are identical. However, the ratio is very often close to unity. For this reason, the curves in Fig. 2.2 are not widely separated. We make the final observation that both channel capacity assignments result in message delays which have poles at $\rho = 1/\bar{n} = 0.764$. This example is pursued further in Sec. 2.2.

Figure 2.2 displays the manner in which the average message delay varies as a function of the network load ρ while a constant total channel

[1] See Eq. (2.9.1) in [46] in which we use the substitutions $r = 1$, $s = 2$, and $a_\nu = \sqrt{\lambda_\nu/\lambda}$.

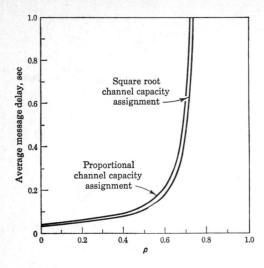

Fig. 2.2. *Comparison of the square root and proportional channel capacity assignments.*

capacity C is maintained. We may also ask for the variation of C which maintains a constant average message delay T, as a function of the data input rate[1] γ/μ. From Eq. (2.5), it is obvious that C must take on values which maintain the quantity $C(1 - \bar{n}\rho)$ constant, i.e.,

$$C(1 - \bar{n}\rho) = K = \frac{\bar{n} \left(\sum\limits_{i=1}^{N} \sqrt{\lambda_i/\lambda} \right)^2}{\mu T}$$

From this we obtain

$$C = K + \frac{\bar{n}\gamma}{\mu} \tag{2.9}$$

Thus Eq. (2.9) shows that there is a simple linear dependence of C on γ/μ which solves the problem stated above.

Constraining the sum of the assigned capacities to be constant implies a special form of system cost. In particular, the implication is that the system cost is represented strictly by the total channel capacity C. As discussed in Sec. 1.8, a more general cost function may be obtained if we consider a function d_i which represents the cost (say in dollars) of supplying one unit of capacity to the ith channel;[2] thus, d_iC_i represents the total cost of assigning the capacity C_i to the ith channel. The optimal channel capacity assignment (namely, that

---

[1] Note that γ/μ is merely the average number of bits that enter the net per second from external sources. We consider μ fixed and γ variable.

[2] For example, d_i may be taken to be proportional to the length of the ith channel.

assignment which minimizes T at a fixed total cost $D = \sum\limits_{i=1}^{N} d_i C_i$) turns out to be

$$C_i = \frac{\lambda_i}{\mu} + \frac{D_e}{d_i} \frac{\sqrt{\lambda_i d_i}}{\sum\limits_{j=1}^{N} \sqrt{\lambda_j d_j}} \tag{2.10}$$

With this assignment, we find that

$$T = \frac{\bar{n} \left(\sum\limits_{i=1}^{N} \sqrt{\lambda_i d_i / \lambda} \right)^2}{\mu D_e} \tag{2.11}$$

where
$$D_e = D - \sum\limits_{j=1}^{N} \frac{\lambda_j d_j}{\mu} \tag{2.12}$$

Equations (2.1) and (2.2) are seen to be the special case of Eqs. (2.10) and (2.11) in which $D = C$ and $d_i = 1$ for all i.

Chapter 5 explores the manner in which message delay is affected when one imposes a priority structure on the set of messages as they pass through a single-channel system. Generally, one breaks the message set into P separate groups, the pth group ($p = 2, 3, \ldots, P$) being given preferential treatment over the $(p - 1)$st group, etc. A newly derived result for a delay-dependent priority system is described in which a message's priority is increased from zero linearly with time, in proportion to a rate assigned to the message's priority group. The usefulness of this priority structure lies in the fact that it provides a number of degrees of freedom with which to manipulate the relative waiting times for each priority group.

An interesting new law of conservation is also proven which constrains the allowed variation in the average waiting times for any one of a wide class of priority structures. Specifically, if we denote by W_p the average time that a message from the pth priority group spends in the queue, then the conservation law states that

$$\sum\limits_{p=1}^{P} \frac{\lambda_p}{\mu_p} W_p = \text{constant with respect to variation of priority structure}$$

where λ_p and $1/\mu_p$ are, respectively, the average arrival rate and average message length for messages from the pth priority group. The analytic expression for this constant is discussed in Chap. 5. As a result of this law, a number of general statements can be made regarding the average waiting times for any priority structure which falls in this class.[1] A priority structure which results in a system of time-

---

[1] See Chap. 5 for an exact description of the class.

shared service is also investigated. This system presents shorter waiting times for "short" messages and longer waiting times for "long" messages; interestingly enough, the critical message length which distinguishes short from long turns out to be the average message length for the case of geometrically distributed message lengths.

Random routing procedures for some specialized nets yield to mathematical analysis and are discussed in Chap. 6. Specifically, a *random routing procedure* is a routing procedure in which the choice of the next node to be visited is made according to some probability distribution over the set of neighboring nodes. The first result obtained therein is the expected number of steps that a message must take (in a net which carries no other traffic) before arriving at its destination for that class of random routing procedures in which the node-to-node transitions are describable by circulant[1] transition matrices. This result exposes the increased number of steps that a message must take in a net with random routing. The next quantity of interest is the expected time that a message spends in the net. The solution for this is presented in Theorem 6.3 (which, once again, makes use of the independence assumption). A quantitative comparison is made between random and fixed routing procedures for identical nets, demonstrating the superiority of the latter as regards message delay.

The last phase of the research describes the results of a large-scale digital simulation of store-and-forward communication nets. The simulation program (written by the author for Lincoln Laboratory's TX-2 computer [52]) is described in Appendix E. Extensive use was made of the simulator in confirming and extending many of the results of this research. For example, it provided a powerful tool for testing the accuracy and suitability of the independence assumption. Furthermore, networks of topological structure identical with those described in Chap. 6 were simulated with fixed routing procedures, and, as predicted, the comparative results indicate that random routing procedures are costly in terms of total traffic handled and message delay. A priority discipline was imposed on the message traffic in some of the runs; results indicate that the conservation law of Chap. 5 holds for nets as well as for a single node.

2.2 Experimental Results

With the background of theoretical results obtained as described above, a careful experimental investigation was carried out (using the

---

[1] A *circulant matrix* is one in which each row is a unit rotation of the row above it [see Eq. (6.2)].

network simulation program) for the purpose of examining the variation of average message delay for different channel capacity assignments, routing procedures, and topologies. A special case of the fixed
cost constraint in Eq. (1.1) was used throughout; that is, we held
constant the total channel capacity assigned to the net, namely
$D = C = \sum_{i=1}^{N} C_i$. These results are presented[1] in Chap. 7. Specifically, it was found that the channel capacity assignment expressed in
Eq. (2.4) was superior to all other assignments tested, not only for
fixed routing procedures (as predicted), but also for a class of alternate
routing procedures.

Furthermore, it was observed that with the square root capacity
assignment, fixed routing was always superior to alternate routing for
the same traffic and the same net. This result becomes more plausible
when one recognizes that alternate routing procedures are designed to
disperse the traffic whenever and wherever it is reasonable to do so.[2]
This is in direct opposition to the result expressed by Eq. (2.3), in
which it is clear that concentrated traffic is to be preferred.

Surprising as this superiority of fixed routing over alternate routing
may be, the statement must be strongly qualified. For, in fact, it was
observed that under a *poor* channel capacity assignment [in violation
of Eq. (2.4)], the performance of alternate routing was itself superior
to that of fixed routing. What we wish to emphasize, then, is that
the simulation results exposed the ability of alternate routing procedures to *adapt* the traffic flow so as to fit the network topology. In
detail, the way in which alternate routing procedures generally operate
is to select an alternative path for a message as it makes its way through
the net whenever its original path becomes blocked beyond some
degree. The effect of such changes is to relieve the heavily loaded
channels (which really have insufficient capacity) and to add to the
load of channels which have an excess capacity. The result is that
the traffic flow is adjusted so that each channel carries an amount of
traffic which is roughly proportional to its capacity. We observed
earlier that a proportional channel capacity assignment was not far
from optimum, and so we may conclude that no significant loss need
be incurred by using alternate routing. A word of caution must be
mentioned: as the network load increases, the use of excessive
alternate routing can be disastrous, since under heavy load conditions
most paths are congested. If alternate routing is given a free hand in

---

[1] The simulation experiments described in this chapter were performed *without*
the use of the independence assumption.

[2] In addition, alternate routing procedures, in general, result in an increased
average path length $(\bar{n})$.

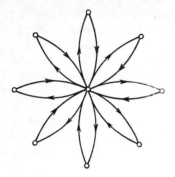

Fig. 2.3.   *Star-net configuration.*

such a situation, most of the messages will be sent on alternative paths, and valuable channel capacity will be wasted on needless side excursions.   The way around this is to enforce stricter and stricter limits on the alternate routing procedure as the network load increases.

This adaptive behavior of alternate routing procedures has considerable significance in the realistic design and operation of a communication net.   Specifically, it is generally true that the actual traffic matrix is not known precisely at the time the network is being designed. Indeed, even if the traffic matrix were known, it is probable that the entries γ_{jk} in this matrix would be time-varying (i.e., different traffic loads would exist at different hours of the day, different days of the week, different seasons of the year, etc.).   In the face of this uncertainty and/or variation, it becomes impossible to calculate the optimum channel capacity assignment from Eq. (2.4), since the numbers λ_i (which are calculable from the γ_{jk} under a fixed routing procedure) are in doubt.   Thus, we propose that one solution to this problem is to use some form of alternate routing which will adapt the actual traffic flow to the network.   Indeed, in such a case, almost any reasonable assignment of capacity will work rather well (see Chap. 7).   Note, however, that a price must be paid for such flexibility, since fixed routing with the square root capacity assignment is itself superior to alternate routing (assuming we have known time-invariant γ_{jk}).

The desirability of a concentrated traffic pattern led to consideration of a special topology, namely, the *star net* shown in Fig. 2.3.   This net has the property that as much traffic as possible is grouped into each channel; the physical constraint here is that the set of origins and destinations (i.e., the traffic matrix) is specified independently of the network design, and so one is forced to have at least one channel leading to and from each node in the net.   The star net yields exactly one channel leading in and out of each node (except, of course, for the central node). The effect of the distribution of traffic λ_i and the average path length $\bar{n}$ on the average message delay in a net with a fixed-routing procedure may be seen in Eq. (2.5). In particular, we note that increased concentration of traffic reduces the expression $\sum_{i=1}^{N} \sqrt{\lambda_i/\lambda}$ [for example see Eq. (2.3)]. On the other hand, we note that T grows without bound as $\rho \to 1/\bar{n}$; recall that $\rho = \gamma/\mu C$ is the ratio of the average arrival rate of bits into the net from external sources to the total capacity of the net. Clearly, a minimum value of $\bar{n}$ is desired. However, it is obvious that the adjustment of λ_i alters the value of $\bar{n}$. In particular, for the star net (which has a maximally concentrated traffic pattern) we observe that $1 < \bar{n} < 2$. If we require a reduced $\bar{n}$, we must add channels to the star net, thus destroying some of the concentration of the traffic. In the limit as $\bar{n} \to 1$, we arrive at the fully connected net which has the smallest possible $\bar{n}$, but also the most dispersed traffic pattern. The trade-off between $\bar{n}$ and traffic concentration depends heavily upon ρ. In particular, we find that, at low network loads, nets topologically similar to the star net are optimum; as the network load increases, we obtain the optimum topology by reducing $\bar{n}$ through the addition of channels; finally, as $\rho \to 1$, we require $\bar{n} = 1$, which results in the fully connected net. In all cases, we use the square root channel capacity assignment with a fixed routing procedure.

Let us illustrate the effect of these topological changes by referring to our telegraph network once again. Figure 2.1 shows our original network with the square root channel capacity assignment adjusted to

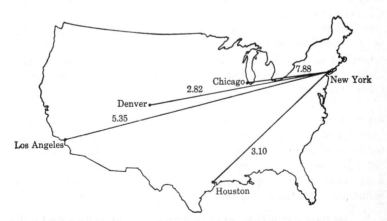

Fig. 2.4. *United States telegraph-network example with a star-net topology.*

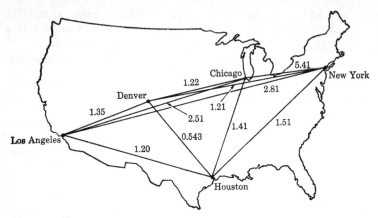

Fig. 2.5. *United States telegraph-network example with a fully connected net topology.*

the traffic matrix τ [see Eq. (2.7)]; this particular traffic matrix represents a network load of $\rho = 0.1$. Figures 2.4 and 2.5 show the topology and the square root channel capacity assignment for the star net and the fully connected net, respectively, for $\rho = 0.1$. We compare the behavior of the average message delays for these three topologies as we allow ρ to vary. The results, calculated from Eq. (2.5), are shown in Fig. 2.6.

These curves bear out our previous statements that only topologies drawn from the optimum sequence described earlier are allowed. Our original net (Fig. 2.1) is not included in this sequence, and so its delay characteristic always lies above the minimum achievable value. In addition, we note that the performance of the star net surpasses that of the fully connected net only for small values of ρ; as ρ increases (beyond 0.6, roughly), the curves cross, and the fully connected net is to be preferred, as predicted. Data obtained from the simulation of other nets may be found in Chap. 7.

A number of interesting results obtained with the help of simulation experiments have been described. These results pertain to the behavior of the average message delay (taken as the measure of performance of the net) with a fixed total channel capacity constraint as the following three design parameters are varied: channel capacity assignment, routing procedure, and topological structure. We now summarize the results of Chap. 7.

1. The square root channel capacity assignment as described in Eq. (2.4) results in superior performance as compared to a number of other channel capacity assignments.

2. The performance of a straightforward fixed routing procedure, with a square root capacity assignment, surpasses that of a simple alternate routing procedure.

3. The alternate routing procedure adapts the internal traffic flow to suit the capacity assignment (i.e., the bulk of the message traffic is routed to the high-capacity channels). This effect is especially noticeable and important in the case of a poor capacity assignment which may come about owing to uncertainty or variation in the applied message traffic.

4. A high degree of nonuniformity in the traffic matrix results in improved performance in the case of a square root channel capacity assignment (owing to a more concentrated traffic pattern).

5. The quantities essential to the determination of the average message delay are the average path length and the degree to which the traffic flow is concentrated. The trade-off between these two quantities allows one to determine the sequence of optimal network topologies, which ranges from nets similar to the star net at small values of network load to the fully connected net as the load approaches unity.

The digital computer as a communication network simulator proved to be an indispensable tool in this investigation. Its single most important contribution was in the simulation of nets *without* the use of the independence assumption. As we shall point out in the next

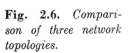

Fig. 2.6. *Comparison of three network topologies.*

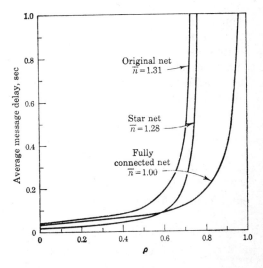

chapter, the mathematical analysis bogs down immediately without this assumption. In order to gain some confidence in its application, we simulated nets both with and without it; as we anticipated, the assumption is quite reasonable. However, we would have been hard pressed to justify its use otherwise. Furthermore, even with the independence assumption, the investigation of nets with alternate-routing procedures would have been unmanageable without the simulation program. A purposeful attempt was made to develop a simulator which would allow an interaction between the computer and the operator. A large measure of success was achieved in that direction, as is described in Appendix E.

The fundamental observation to be made in this regard is that the intelligent use of simulators allows one to investigate realistic physical systems (or their mathematical representations) whose complexity is such that full mathematical treatment is hopeless. There is a vast domain of problems to which careful simulation techniques may profitably be applied. The results of such an investigation enhance one's understanding of the physical system; this then often enables one to make appropriate simplifying assumptions about the original system and subsequently to carry out the mathematics. Also, certain aspects of the asymptotic behavior of the system are often revealed in a simulation; such behavior may usually be examined without difficulty in the original physical system.

We shall close this chapter by elaborating somewhat on system cost functions for communication nets. The particular function for which we have solutions is described by Eq. (1.1). This function allows us to take into account the length of a channel, the terrain over which the channel may lie, etc. In the examples presented in this chapter, we have assumed for simplicity that the cost of a channel is equal merely to its capacity. Furthermore, we have omitted the consideration of unreliable nodes and links. A most interesting problem which incorporates the extension to unreliability as well as to the more general cost function described by Eq. (1.1) may be expressed as

$$\text{Minimize} \qquad AT + BU$$

$$\left\{ \begin{array}{l} \text{Capacity assignment,} \\ \text{routing procedure,} \\ \text{priority discipline,} \\ \text{topology} \end{array} \right\}$$

subject to the fixed cost constraint

$$D = \sum_{i=1}^{N} d_i C_i$$

where A and B represent one's dislike for a unit of message delay T and unreliability U, respectively.   Of course, the function U must be an appropriately defined measure of the unreliability.   Another interesting extension which occurs in certain other existing systems is one in which costs are not linearly related to capacity.   More general cost functions and problem statements can obviously be written (and each change in cost will result in different optimal solutions); however, the form shown above represents a sufficiently difficult challenge for the present.

chapter 3

The Problems of an Exact Mathematical Solution to the General Communication Net

3.1 Discussion of the General Problem

We have before us the task of supplying answers to the various questions posed by the designer in Sec. 1.4. We therefore require a mathematical description of the behavior of the message delay as the design parameters[1] are varied. One way in which such a description can be obtained is by consideration of an appropriate set of *state variables*. Such a set of variables must satisfy two conditions. First, it must include (explicitly or implicitly) those quantities which are of interest, e.g., the message delay. Second, the set must be complete, or closed, such that knowledge of the state variables at time t and knowledge of all message arrivals from sources external to the net in the closed interval (t,t') are sufficient to specify uniquely the state

---

[1] The design parameters under consideration are capacity assignment to each channel, topological structure, routing procedure, priority discipline, total traffic handled, and total system cost.

variables at time $t' \geq t$. This second condition describes the markovian property.

One set of state variables which satisfies these conditions will now be defined. We consider a communication net with N nodes and M one-way channels, under the assumptions described in Sec. 1.6. It is clear that the state of the net at any instant of time must include detailed information as to the number of messages on each queue, the length of each message, and the time required to complete the transmission in progress on each channel. Furthermore, we assume that each message is labeled with an origin, a destination, and a priority. Accordingly, we define

C_i = capacity of ith channel, bits/sec

n_i = number of messages waiting for (or being transmitted on) ith channel

γ_i = average arrival rate of messages from external sources to ith channel

$v_{in}^{(x_{in}, y_{in}, z_{in})}$ = message length,[1] in bits, of nth message waiting for (or being transmitted on) ith channel, whose origin, destination, and priority are x_{in}, y_{in}, and z_{in}, respectively; for conciseness of notation, let symbol v_{in} denote this length

V_{n_i} = set of numbers $(v_{i1}, v_{i2}, \ldots, v_{in_i})$

r_i = time remaining to complete transmission in progress on ith channel

R = set of numbers $(r_1, r_2, \ldots, r_M)$

$R + dt$ = set of numbers $(r_1 + dt, r_2 + dt, \ldots, r_M + dt)$

where $i = 1, 2, \ldots, M$ and $n = 1, 2, \ldots, n_i$

The *state* of the net at any time t may be completely described by the set of variables

$$S = (n_1, n_2, \ldots, n_M, V_{n_1}, V_{n_2}, \ldots, V_{n_M}, R)$$

Clearly, all these quantities are functions of time. This state description is of unbounded dimensionality, since the variables n_i are unbounded. Furthermore, all these variables are necessary in order that the state description be complete.

For each state S and each time t a probability density function $p_t(S)$ is associated with the probability that the net will be found in state S at time t. In general, one desires an explicit solution for the function $p_t(S)$. To date, this problem remains unsolved. We shall carry out

---

[1] Recall that the distribution of message lengths is exponential with mean length $1/\mu$.

a portion of the analysis in an effort to indicate the source of the difficulty. Let us set up the equations under the conditions

$$n_i > 1$$
$$0 < r_i < \frac{v_{i1}}{C_i}$$

The first condition, $n_i > 1$, is included for convenience at this point. The end points zero and v_{i1}/C_i are excluded from the allowed range of r_i in order to eliminate temporarily from discussion any consideration of internal message arrivals.[1] For this case, we immediately write

$$p_{t+dt}(n_1, n_2, \ldots, n_M, V_{n_1}, V_{n_2}, \ldots, V_{n_M}, R)$$
$$= p_t(n_1, n_2, \ldots, n_M, V_{n_1}, V_{n_2}, \ldots, V_{n_M}, R + dt) \left(1 - \sum_{i=1}^{M} \gamma_i \, dt\right)$$
$$+ \sum_{i=1}^{M} \gamma_i' \, dt \, \mu e^{-\mu v_{in_i}} p_t(n_1, n_2, \ldots, n_i - 1, \ldots, n_M,$$
$$V_{n_1}, V_{n_2}, \ldots, V_{n_i - 1}, \ldots, V_{n_M}, R + dt)$$

where γ_i' represents that portion of γ_i which has the x_{in_i}, y_{in_i}, and z_{in_i} which correspond to v_{in_i}. This leads us to the following partial-differential difference equation:

$$\frac{\partial p_t}{\partial t}(n_1, n_2, \ldots, n_M, V_{n_1}, V_{n_2}, \ldots, V_{n_M}, R) \cdot$$
$$- \sum_{i=1}^{M} \frac{\partial p_t}{\partial r_i}(n_1, n_2, \ldots, n_M, V_{n_1}, V_{n_2}, \ldots, V_{n_M}, R) =$$
$$\sum_{i=1}^{M} \gamma_i' \mu e^{-\mu v_{in_i}} p_t(n_1, n_2, \ldots, n_i - 1, \ldots, n_M,$$
$$V_{n_1}, V_{n_2}, \ldots, V_{n_i - 1}, \ldots, V_{n_M}, R)$$
$$- p_t(n_1, n_2, \ldots, n_M, V_{n_1}, V_{n_2}, \ldots, V_{n_M}, R) \sum_{i=1}^{M} \gamma_i$$

where $n_i > 1$ and $0 < r_i < v_{i1}/C_i$. The equations involving r_i at its end points force one to consider internal message arrivals and thus become considerably more complex. One must then include the rules of the routing procedure in determining which transitions occur. The task of solving this set of partial-differential difference equations is formidable, and no solution has yet been found.

---

[1] An internal message arrival occurs when a message completes its transmission between two nodes internal to the net (as opposed to an external message arrival which occurs when a message arrives at its origin from a source external to the net).

The complexity of the state description is due in part to the constraint that each message, upon entering the net, has a permanent length assigned to it. The message maintains this same length as it travels through the net. This clearly necessitates the inclusion of the variables V_{n_i} in the state description. The identification of a permanent length with each message not only increases the dimensionality of the state description, but also complicates the stochastic behavior of the net by introducing a dependence among some of the random variables which describe the behavior of the net.

In particular, if we consider two successive messages arriving at node i from some other node internal to the net, then the interarrival time between these messages is not independent of the message length of the second of the two messages.[1] More specifically, let us now derive the joint probability density function $p(v_n, a_{2n})$ for the simple two-node tandem net shown in Fig. 3.1, where we define

v_n = message length of nth message, bits
a_{in} = time between arrivals of $(n - 1)$st and nth messages at node i

For convenience, we take $C_1 = 1$ bit/sec. Clearly this represents no loss of generality.

By the assumptions of Sec. 1.6, we recall that both a_{1n} and v_n are described by the following exponential probability density functions

$$p(a_{1n}) = \gamma e^{-\gamma a_{1n}}$$
$$p(v_n) = \mu e^{-\mu v_n} \tag{3.1}$$

Further, for our immediate purposes, it is convenient to assume that all messages originate at node 1 and are required to pass through nodes 1 and 2.

Observe that channel C_1 fits the classical queueing-theory model of a single exponential channel system as described in Appendix A; hence, all results from that appendix apply. Since Theorem A.1 (due to Burke) holds, and since the interdeparture times for messages leaving node 1 are, by definition, identical to the interarrival times for messages entering node 2, we see that

$$p(a_{2n}) = \gamma e^{-\gamma a_{2n}} \tag{3.2}$$

The nth message leaving node 1 is either (1) separated by a time gap g_n from the $(n - 1)$st message or (2) transmitted immediately after the

---

[1] Of course, this independence exists by assumption for messages which arrive from an external source.

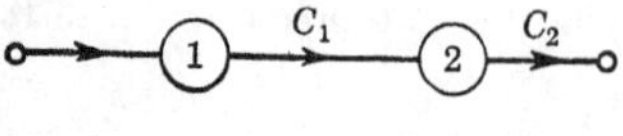

Fig. 3.1.   *Two-node tandem net.*

$(n - 1)$st message is finished (see Fig. 3.2). Case 1 occurs only if the first node emptied while awaiting the nth message, and this happens with probability $1 - \rho$ [see Eq. (A.3) in Appendix A], where

$$\rho = \frac{\gamma}{\mu C_1} = \frac{\gamma}{\mu}$$

Case 2 occurs if the first node is busy when the nth message arrives; this has probability ρ. Thus[1]

$$p(v_n, a_{2n}) = \rho P_r[v_n, a_{2n}|\text{node 1 busy}] + (1 - \rho)P_r[v_n, a_{2n}|\text{node 1 empty}] \tag{3.3}$$

Clearly,[2] $P_r[v_n, a_{2n}|\text{node 1 busy}] = P_r[v_n = a_{2n}] = \mu e^{-\mu v_n} u_o(a_{2n} - v_n)$
and

$$P_r[v_n, a_{2n}|\text{node 1 empty}] = P_r[a_{2n}|v_n, \text{node 1 empty}]P_r[v_n|\text{node 1 empty}]$$
$$= P_r[g_n = a_{2n} - v_n]P_r[v_n|\text{node 1 empty}]$$

Due to the memoryless property of an exponential distribution (see the discussion on page 150), g_n is also distributed according to Eq. (3.2). Thus

$$P_r[v_n, a_{2n}|\text{node 1 empty}] = \gamma e^{-\gamma(a_{2n} - v_n)} \mu e^{-\mu v_n}$$

Therefore, Eq. (3.3) becomes

$$p(v_n, a_{2n}) = \gamma e^{-\mu v_n} u_o(a_{2n} - v_n) + \gamma(\mu - \gamma)e^{-\gamma(a_{2n} - v_n) - \mu v_n} \tag{3.4}$$

This last equation gives the desired joint probability density function of v_n and a_{2n}.

---

[1] The notation $P_r[x]$ is to be interpreted as "the probability of x."
[2] $u_o(x)$ is the unit impulse function.

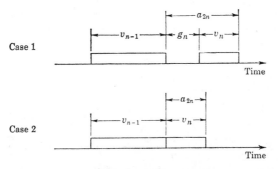

Fig. 3.2. *Adjacent messages leaving node 1.*

Upon comparing Eqs. (3.1), (3.2), and (3.4), we see that

$$p(v_n, a_{2n}) \neq p(v_n)p(a_{2n})$$

which, by definition, illustrates a lack of independence between v_n and a_{2n}. This dependence is a source of great complication in the exact mathematical analysis of the general net; indeed, no general solution has been obtained.

3.2   The Tandem Net

If we consider the tandem net as shown in Fig. 3.3, we simplify the general problem somewhat.[1]   In particular, we remove from consideration the questions of origin, destination, and routing procedure, since all messages (by assumption) originate at node 1, and are routed successively through nodes 1, 2, . . . , K. We do, however, retain the dependency between the interarrival time and the lengths of messages.   In addition, since we have only one external input (at node 1), the complete course of a message can be calculated deterministically as soon as that message arrives at node 1.   That is, we can state exactly how long that message will spend in each node k ($k = 1$, 2, . . . , K).   Nevertheless, the complete mathematical solution, even for this simplified net, evades us.   We have been able to obtain some partial results, which we now proceed to describe.

Both v_n and a_{in} retain their definitions from Sec. 3.1.   We introduce additional notation w_{kn} and g_{kn} as follows:

w_{kn} = total time that nth message spends in node k

g_{kn} = time that kth node remains idle while awaiting arrival of nth message

In Fig. 3.4 are shown graphically the histories of seven messages as they pass through the first three tandem nodes.   The function plotted is the *total unfinished work* $U_k(t)$ in the kth node; it represents the total number of seconds that will elapse before the kth node empties if no other messages enter this node after time t.   We remind the reader at this point that a message is not considered to have arrived at a node until the entire message is received at that node (in spite of the fact that message reception takes place sequentially in time).   For

[1] Note that $C_k = C$ for $k = 1, 2, . . . , K$.   We take $C = 1$ for convenience.

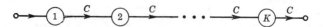

Fig. 3.3.   *Tandem net with K nodes.*

simplicity, the fourth message is the only one labeled in the second and third nodes.

It is clear that if the nth message arrives at node k at time t, it adds v_n to $U_k(t^-)$. That is,[1]

$$U_k(t^+) = U_k(t^-) + v_n \qquad (3.5)$$

But
$$w_{kn} = U_k(t^+)$$

Now, if $g_{kn} > 0$, which means that the kth node was idle at time t^-, then $U_k(t^-) = 0$, and so

$$w_{kn} = v_n \qquad \text{for } g_{kn} > 0$$

If, on the other hand, $g_{kn} = 0$, then $U_k(t^-)$ was still positive and decreasing at a rate of 1 sec/sec; in fact, since exactly a_{kn} seconds had elapsed since the $(n - 1)$st message arrived,

$$U_k(t^-) = w_{k,n-1} - a_{kn}$$

[1] t^- is defined as $t - dt$, and t^+ as $t + dt$.

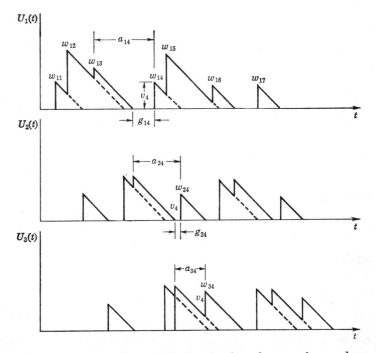

Fig. 3.4. *Example of $U_k(t)$ for the first three tandem nodes.*

and so, from Eq. (3.5), we obtain

$$w_{kn} = w_{k,n-1} - a_{kn} + v_n \qquad \text{for } g_{kn} = 0 \tag{3.6}$$

Note that, for $k \geq 2$, the interarrival time a_{kn} is made up of the transmission time out of the $(k-1)$st node for the nth message, plus any time that the $(k-1)$st node spent idle awaiting the arrival of the nth message. That is,

$$a_{kn} = v_n + g_{k-1,n} \tag{3.7}$$

Therefore, Eq. (3.6) may be written as

$$w_{kn} = w_{k,n-1} - g_{k-1,n} \qquad \text{for } g_{kn} = 0 \text{ and } k \geq 2$$

Furthermore, it is clear that the maximum amount by which the kth node can reduce its unfinished work in the time between the arrivals of the $(n-1)$st and nth messages is a_{kn}. Thus, if $a_{kn} \leq w_{k,n-1}$, then the idle time g_{kn} in node k before the nth arrival will be zero. If $a_{kn} \geq w_{k,n-1}$, then $g_{kn} = a_{kn} - w_{k,n-1}$. Summarizing the results for the tandem net so far, we have

$$w_{1n} = \begin{cases} w_{1,n-1} - a_{1n} + v_n & \text{for } g_{1n} = 0 \\ v_n & \text{for } g_{1n} > 0 \end{cases} \tag{3.8}$$

$$w_{kn} = \begin{cases} w_{k,n-1} - g_{k-1,n} & \text{for } g_{kn} = 0 \text{ and } k \geq 2 \\ v_n & \text{for } g_{kn} > 0 \text{ and } k \geq 2 \end{cases} \tag{3.9}$$

$$g_{kn} = \begin{cases} 0 & \text{for } a_{kn} \leq w_{k,n-1} \\ a_{kn} - w_{k,n-1} & \text{for } a_{kn} \geq w_{k,n-1} \end{cases} \tag{3.10}$$

We now proceed to derive an expression for $\Pr[w_{kn} \leq W]$ for $k \geq 2$. From Eqs. (3.9) and (3.10), we immediately obtain

$$\Pr[w_{kn} \leq W] = \Pr[w_{k,n-1} - g_{k-1,n} \leq W, \, a_{kn} \leq w_{k,n-1}] \\ + \Pr[v_n \leq W, \, a_{kn} \geq w_{k,n-1}]$$

By use of Eq. (3.7), we obtain

$$\Pr[w_{kn} \leq W] = \Pr[v_n \leq w_{k,n-1} - g_{k-1,n} \leq W] \\ + \Pr[w_{k,n-1} - g_{k-1,n} \leq v_n \leq W]$$

Defining $x_{kn} = w_{k,n-1} - g_{k-1,n}$, we observe that the above equation integrates the probability over the region in the product space of x_{kn} and v_n such that $x_{kn} \leq W$ and $v_n \leq W$. Thus, we finally obtain

$$\Pr[w_{kn} \leq W] = \Pr[v_n \leq W, \, w_{k,n-1} \leq W + g_{k-1,n}] \tag{3.11}$$

This is an interesting and general result for the tandem net with $k \geq 2$.

Let us now consider the limiting behavior of the message traffic leaving the Kth node, as $K \to \infty$. We introduce a *busy period* for the kth node as being an interval of time during which the kth node is continuously transmitting messages. We shall also refer to the mth busy period for node k as the mth occurrence of a busy period for that node. Define

$$L_{mk} = \text{length of message which initiates } m\text{th busy period in}$$
$$\text{node } k, \text{ bits (or seconds, since } C = 1)$$

$$T_{mk} = \text{time between starts of } m\text{th and } (m - 1)\text{st busy periods}$$
$$\text{in node } k$$

$$L_m = \lim_{K \to \infty} L_{mK}$$

$$T_m = \lim_{K \to \infty} T_{mK}$$

$$p(i|L_m) = \lim_{K \to \infty} \text{Pr}[i \text{ messages in } m\text{th busy period of } K\text{th node,}$$
$$\text{given } L_{mK}]$$

$$p'(L|L_m) = \lim_{K \to \infty} \text{Pr}[L \text{ given } L_{mK}, \text{ where } L + L_{mK} = \text{total length of}$$
$$m\text{th busy period in node } K]$$

With these definitions, we now state:

Theorem 3.1

As $K \to \infty$, the limiting form of the message traffic as it leaves the Kth node behaves as follows:

1. All messages in the mth busy period spend exactly L_m seconds in the Kth node, where $L_m = L_{m2}$
2. $L_m \geq L_{m-1}$ for $m = 2, 3, \ldots$
3. $T_m = \begin{cases} T_{m1} & \text{if } L_m = L_{m-1} \\ \infty & \text{if } L_m > L_{m-1} \end{cases}$
4. $p(i|L_m) = e^{-\mu L_m}(1 - e^{-\mu L_m})^{i-1}$
5. $p'(L|L_m) = \mu(1 - e^{-\mu L_m})e^{-\mu(L_m + Le^{-\mu L_m})} + e^{-\mu L_m}u_o(L)$

PROOF: Let us consider the mth busy period in node k. We assume that there are m_k messages included in this period. Note that $g_{kn} = 0$ for all messages grouped in a busy period, except for the message which initiates the busy period. With this condition, Eq. (3.9) states that, for $k \geq 2$,

$$w_{kn} = w_{k,n-1} - g_{k-1,n}$$

and so we see that, for $k \geq 2$,

$$w_{k1} \geq w_{k2} \geq \cdots \geq w_{km_k} \tag{3.12}$$

Now, since $g_{kn} = 0$ within the busy period, we obtain, from Eq. (3.7),

$$a_{k+1,n} = v_n$$

Clearly, since w_{kn} is made up of transmission time v_n plus queueing time,

$$v_n \leq w_{k+1,n}$$

Thus

$$a_{k+1,n} \leq w_{k+1,n}$$

Applying Eq. (3.12), we obtain

$$a_{k+1,n} \leq w_{k+1,n} \leq w_{k+1,n-1}$$

From this and Eq. (3.10), we find that $g_{k+1,n} = 0$. Applying this to Eq. (3.9), and recalling that $g_{kn} = 0$ within the busy period, we obtain, for $k \geq 2$,

$$w_{k+1,n} = w_{k+1,n-1} \tag{3.13}$$

In summary, Eq. (3.12) states that all messages in a busy period have monotonically decreasing waiting times. Further, Eq. (3.13) states that after messages which are grouped into a busy period pass through the next node, they all have *identical* waiting times. As the group passes through this node, new messages may be added to the busy period; the waiting time for these need only obey Eq. (3.12). However, only messages with $v_n \leq w_{k1}$ can possibly join this group in the busy period [because of Eq. (3.12)]. Thus, after passing through an unbounded number of nodes, all messages in a busy period will have identical waiting times. Thus, $L_m = L_{m2}$. This proves statement 1 of the theorem. Reference to Fig. 3.4 will aid the reader in his understanding of this portion of the proof.

We now observe that in passing through a node the initiating message of the mth busy period (hereafter referred to as the mth *group leader*) gets delayed by exactly L_m seconds (its own transmission time) for $k \geq 2$, and so

$$T_{mk} = T_{m,k-1} + L_m - L_{m-1}$$
or
$$T_{mk} = (k-1)(L_m - L_{m-1}) + T_{m1}$$

Now, if $L_m - L_{m-1} < 0$, then as k increases, T_{mk} will eventually go negative; but this implies that the $(m-1)$st and mth busy periods have coalesced into one large busy period. If $L_m - L_{m-1} > 0$, then as k increases, $T_{mk} \to \infty$, and for $L_m = L_{m-1}$, $T_{mk} = T_{m1}$. From this argument, we see that as $k \to \infty$, $L_m \geq L_{m-1}$, where the subscript m refers only to distinct busy periods (by definition). We have thus established statements 2 and 3 of the theorem.

Furthermore, we see that, as $k \to \infty$, the mth busy period will contain only those messages which arrived at node 1 after the mth group leader and before the $(m + 1)$st group leader. This also implies, by Eq. (3.12), that $v_n \leq L_m$ for those n included in the mth busy period. Since the arrival times at node 1 are chosen independently, we easily calculate $p(i|L_m)$ as

$$p(i|L_m) = \mathrm{P_r}[i - 1 \text{ messages arrive, each with } v_n < L_m, \; n = 1, 2,$$
$$\ldots , \, i - 1] \, \mathrm{P_r}[i\text{th message has length } v_i \geq L_m]$$
$$= \mathrm{P_r}[\text{message length} \geq L_m] \, \mathrm{P_r}[\text{message length} < L_m]^{i-1}$$

Owing to the exponential distribution of message lengths,

$$p(i|L_m) = e^{-\mu L_m}(1 - e^{-\mu L_m})^{i-1} \qquad i \geq 1$$

which proves statement 4 of the theorem.

Proceeding with $p'(L|L_m)$, we note immediately that

$$p'(L|L_m) = p(1|L_m)u_0(L)$$
$$+ \sum_{n=2}^{\infty} p(n|L_m)\mathrm{P_r}[\text{sum of } n - 1 \text{ message lengths} = L]$$

Now, since all message lengths are independent random variables, the probability density of the sum of these $n - 1$ random variables is merely the convolution of their individual probability density functions. Performing this convolution on the exponentially distributed lengths, we obtain

$$p'(L|L_m) = p(1|L_m)u_0(L) + \sum_{n=2}^{\infty} p(n|L_m) \frac{\mu(\mu L)^{n-2}e^{-\mu L}}{(n - 2)!}$$

Substituting for $p(n|L_m)$ from statement 4 of the theorem and performing the indicated summation, we arrive at the expression given by statement 5. This completes the proof of Theorem 3.1.

The interesting results which describe the behavior of message traffic in a tandem net are given by Eqs. (3.8) to (3.11) and Theorem 3.1.

3.3 The Two-node Tandem Net

If we limit ourselves to the study of a tandem net with $K = 2$ and allow the possibility of $C_1 \neq C_2$, we find that we are able to carry the analysis further (although not to completion). Specifically, we are able to derive a functional equation for the Laplace transform of the joint distribution of a message's length, its time spent on the queue in the first node, and its time spent on the queue in the second node.

From this, we obtain a similar transform expression for the marginal distribution of the queueing time in the second node.

We first introduce notation suitable for this two-node case:

x = queueing time for $(n - 1)$st message in node 1
y = queueing time for $(n - 1)$st message in node 2
q = queueing time for nth message in node 1
r = queueing time for nth message in node 2
u = length of $(n - 1)$st message, bits
v = length of nth message, bits
a = interarrival time for $(n - 1)$st and nth messages at first node

We are interested in obtaining an expression for queueing time in the second node, since this is the node that is supplied with the dependent traffic.   We know the waiting and queueing time for the first node, since it satisfies the conditions of a single exponential channel (i.e., it is fed from an external source with the suitable independence between message lengths and interarrival times of messages).   In solving for the queueing time in the second node, we are forced to consider the joint distribution of the triplet (x,y,u), since it is this distribution which appears in our probability expressions below.   We proceed by expressing the joint distribution of (q,r,v) in terms of the distribution of (x,y,u); we then take the limit of these expressions as $n \rightarrow \infty$.   This results in an integral equation whose Laplace transform we then obtain.

We know that the marginal distribution of all message lengths (for example, u and v) is exponential with mean length $1/\mu$; further, the interarrival times a are also distributed exponentially with mean length $1/\lambda$.   We now define a set of probability expressions

$$P_1 = \Pr[q = 0, r = 0, v \leq V]$$
$$P_2 = \Pr[0 < q \leq Q, r = 0, v \leq V]$$
$$P_3 = \Pr[q = 0, 0 < r \leq R, v \leq V]$$
$$P_4 = \Pr[0 < q \leq Q, 0 < r \leq R, v \leq V]$$
$$P = \Pr[0 \leq q \leq Q, 0 \leq r \leq R, v \leq V]$$

From these definitions, it is obvious that

$$P = P_1 + P_2 + P_3 + P_4$$

$P = P(Q,R,V)$ is the three-dimensional cumulative probability function in which we are interested.   Corresponding to P there may be defined, with the help of impulse functions, the probability density function

$$p(Q,R,V) = \frac{\partial^3 P}{\partial Q\, \partial R\, \partial V} (Q,R,V)$$

Similarly, we define

$$p_1(V) = \frac{\partial P_1}{\partial V}(V)$$

$$p_2(Q,V) = \frac{\partial^2 P_2}{\partial Q\, \partial V}(Q,V)$$

$$p_3(R,V) = \frac{\partial^2 P_3}{\partial R\, \partial V}(R,V)$$

$$p_4(Q,R,V) = \frac{\partial^3 P_4}{\partial Q\, \partial R\, \partial V}(Q,R,V)$$

It is then clear that

$$p(Q,R,V) = p_1(V)u_o(Q)u_o(R) + p_2(Q,V)u_0(R) + p_3(R,V)u_o(Q) + p_4(Q,R,V)$$

For conciseness, we define the following quantities:

$$V_i = \frac{V}{C_i}$$

$$D = x + y + \frac{u}{C_1} + \frac{u}{C_2}$$

$$E = (V_1 - y)C_2$$

$$F = (Q - x)C_1$$

$$G = \begin{cases} 0 & Q \leq V_1 \dfrac{C_2}{C_1} \\[2mm] Q - V_1 \dfrac{C_2}{C_1} & Q \geq V_1 \dfrac{C_2}{C_1} \end{cases}$$

$$H = \begin{cases} 0 & Q \leq \dfrac{C_2}{C_1}(V_1 + R) \\[2mm] Q - \dfrac{C_2}{C_1}(V_1 + R) & Q \geq \dfrac{C_2}{C_1}(V_1 + R) \end{cases}$$

Omitting the lengthy arguments involved, we present the derived expressions for p_i ($i = 1, 2, 3, 4$):

$$
\begin{aligned}
p_1(V) =\ & \int_{x=0}^{\infty}\int_{y=0}^{V_1}\int_{u=0}^{E} \mu p(x,y,u)e^{-\lambda(x+u/C_1)-\mu V}\, dx\, dy\, du \\
& + \int_{x=0}^{\infty}\int_{y=0}^{V_1}\int_{u=E}^{\infty} \mu p(x,y,u)e^{-\lambda(D-V_1)-\mu V}\, dx\, dy\, du \\
& + \int_{x=0}^{\infty}\int_{y=V_1}^{\infty}\int_{u=0}^{\infty} \mu p(x,y,u)e^{-\lambda(D-V_1)-\mu V}\, dx\, dy\, du \\
p_2(Q,V) =\ & \int_{x=G}^{Q}\int_{y=0}^{V_1-F/C_2}\int_{u=F}^{E} \mu\lambda p(x,y,u)e^{-\lambda(x+u/C_1-Q)-\mu V}\, dx\, dy\, du \\
& + \int_{x=Q}^{\infty}\int_{y=0}^{V_1}\int_{u=0}^{E} \mu\lambda p(x,y,u)e^{-\lambda(x+u/C_1-Q)-\mu V}\, dx\, dy\, du
\end{aligned}
$$

$$p_3(R,V) = \int_{x=0}^{\infty} \int_{y=0}^{V_1+R} \int_{u=(V_1+R-y)C_2}^{\infty} \mu\lambda p(x,y,u)e^{-\lambda(D-V_1-R)-\mu V}\,dx\,dy\,du$$

$$+ \int_{x=0}^{\infty} \int_{y=V_1+R}^{\infty} \int_{u=0}^{\infty} \mu\lambda p(x,y,u)e^{-\lambda(D-V_1-R)-\mu V}\,dx\,dy\,du$$

$$p_4(Q,R,V) = \int_{x=H}^{Q} \int_{y=0}^{V_1+R-F/C_2} \mu\lambda C_2 p[x,\,y,\,u = C_2(V_1+R-y)]$$

$$e^{-\lambda(D-V_1-Q-R)-\mu V}\,dx\,dy$$

$$+ \int_{x=Q}^{\infty} \int_{y=0}^{V_1+R} \mu\lambda C_2 p[x,\,y,\,u = C_2(V_1+R-y)]$$

$$e^{-\lambda(D-V_1-Q-R)-\mu V}\,dx\,dy$$

We now define the three-dimensional Laplace transform

$$L(s_1,s_2,s_3) = \int_{Q=0}^{\infty} \int_{R=0}^{\infty} \int_{V=0}^{\infty} p(Q,R,V)e^{-s_1 Q-s_2 R-s_3 V}\,dQ\,dR\,dV$$

After taking transforms of the above expressions and collecting terms, we finally obtain the following expression for $L(s_1,s_2,s_3)$:

$$L(s_1,s_2,s_3) = L\left(\lambda,\lambda,\frac{\lambda}{C_1}+\frac{\lambda}{C_2}\right)\frac{\mu s_2}{(s_2-\lambda)(\mu+s_3-\lambda/C_1)}$$

$$- L\left(\lambda,(\mu+s_3)C_1,\frac{\lambda}{C_1}+(\mu+s_3)\frac{C_1}{C_2}\right)$$

$$\frac{\mu\lambda s_2(\mu+s_3-s_1/C_1)}{C_1(\mu+s_3)(s_1-\lambda)(\mu+s_3-\lambda/C_1)(\mu+s_3-s_2/C_1)}$$

$$+ L\left(\lambda,s_2,\frac{\lambda}{C_1}+\frac{s_2}{C_2}\right)\frac{\mu\lambda(s_2-s_1)}{(\mu+s_3-s_2/C_1)(s_1-\lambda)(s_2-\lambda)}$$

$$- L\left(s_1,s_2,\frac{s_1}{C_1}+\frac{s_2}{C_2}\right)\frac{\mu\lambda}{(s_1-\lambda)(\mu+s_3-s_2/C_1)}$$

$$+ L\left(s_1,(\mu+s_3)C_1,\frac{s_1}{C_1}+(\mu+s_3)\frac{C_1}{C_2}\right)$$

$$\frac{\mu\lambda s_2}{C_1(s_1-\lambda)(\mu+s_3-s_2/C_1)(\mu+s_3)} \tag{3.14}$$

Equation (3.14) represents the extent to which the solution has been carried.   Note that for $s_1 = s_2 = 0$ the proper marginal distribution for the message length is obtained (after the inverse transform is taken, of course).   Furthermore, for $s_2 = s_3 = 0$, we obtain the expression for the transform of the marginal distribution of the queueing time in the first node.[1]

---

[1] Note that after multiplying this by the transform of the service-time distribution, we obtain the transform of the total time spent in the first node.   Inverting this product, we arrive at an expression which agrees with Eq. (A.5) in Appendix A.

With $s_1 = s_3 = 0$, we obtain an expression for the Laplace transform of the queueing time in the second node, as follows:

$$L(0,s_2,0) = L\left(\lambda, \lambda, \frac{\lambda}{C_1} + \frac{\lambda}{C_2}\right) \frac{\mu s_2}{(s_2 - \lambda)(\mu - \lambda/C_1)}$$

$$+ \frac{\mu s_2}{\mu - s_2/C_1}\left[\frac{L(\lambda, \mu C_1, \lambda/C_1 + \mu C_1/C_2)}{\mu C_1 - \lambda} - \frac{L(\lambda, s_2, \lambda/C_1 + s_2/C_2)}{s_2 - \lambda}\right]$$

$$+ \frac{1}{\mu - s_2/C_2}\left[\mu L(0, s_2, s_2/C_2) - \frac{s_2}{C_1} L(0, \mu C_1, \mu C_1/C_2)\right] \quad (3.15)$$

This functional equation has not been solved.

It is interesting to note that even the solution for the marginal distribution of the queueing time in the second node escapes us. Furthermore, we observe that the case under consideration is the simplest one in which the effect of the dependency between the interarrival times and lengths of messages may be analyzed, and yet a solution was not obtained.

3.4 The Independence Assumption

We recognize that the source of difficulty in solving the general net (or even the simpler tandem net) lies in the assignment of a permanent length to each message. This permanent assignment gives rise to a dependency [see, for example, Eq. (3.4)] between the interarrival times and lengths of adjacent messages as they travel within the net. Indeed, as we shall see below, the elimination of this dependency simplifies the mathematics considerably.

Recall the assumption of independence between the arrival time and length of a message as it enters the net from an external source. We stated that this assumption was quite accurate in describing the externally applied traffic for some communication nets. We may now inquire as to what properties of the external traffic bring about this independence. The answer is straightforward and may be found by observing that the external message source consists of a large number of subscribers (people), each individually generating messages (telegrams) at a relatively small rate. The interarrival times and lengths of messages generated by any individual are indeed dependent in a manner not unlike that expressed by Eq. (3.4).[1] However, the *collective* interarrival times and lengths of messages generated by the

---

[1] That is, any individual requires a finite amount of time to generate a message, and the length of this interval of time is strongly dependent upon the length of the message.

entire population of subscribers exhibit an independence, since the length of a message generated by any particular subscriber is completely independent of the arrival times of messages generated by the other subscribers.

A similar situation exists for the internal traffic of many practical store-and-forward communication nets. That is, there is, in general, more than one channel delivering messages into any particular node (in addition to those messages arriving from the external source feeding this node). Furthermore, there is, in general, more than one channel transmitting messages out of this node (in addition to the "virtual" channel which removes those messages which had this node as a final destination). Fortunately (for analytical purposes), this multiplicity of paths in and out of each node considerably reduces the dependency between interarrival times and lengths of messages as they enter various channels (or queues) within the net. We offer evidence of this essential independence with the experimental results described in the next section.

If, indeed, this assumption of independence describes the general network behavior to a fair degree of accuracy, and if, at the same time, this assumption simplifies the mathematics, then we have good reason to accept it. Specifically, one way in which we can introduce independence into the mathematical description is to make the following assumption:

The Independence Assumption

Each time a message is received at a node within the net, a new length v is chosen for this message from the following probability density function:

$$p(v) = \mu e^{-\mu v}$$

It is clear that this assumption does not correspond to the actual situation in any practical communication net. Nevertheless, its mathematical consequences result in a model which accurately describes the behavior of the message delay in many real nets. We offer evidence of this in the next section.

3.5 The Effect of the Independence Assumption

Although the strict mathematical approach to the general net has resulted in, at most, limited analytical results, we still require an

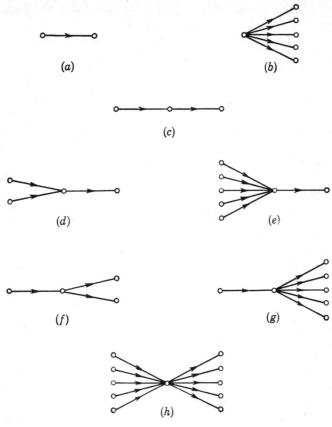

Fig. 3.5. *Simulated nets for studying internal network traffic.*

answer of some kind to the problem of a general configuration. We have presented a loose heuristic argument as to why the independence assumption represents a useful simplification of the problem. Up to now, we have offered no substantial evidence of the accuracy of this assumption. There is at our disposal a powerful tool with which to test its accuracy, namely, a digital-computer simulation program. The simulator is described in Appendix E.

We first present simulator results demonstrating the effect on message delay of the introduction of additional channels leading into and emanating from a node. The network configurations which were simulated are shown in Fig. 3.5. The histogram of message delay and the average message delay were the quantities obtained from the simulation.

Note that in configurations c to h there are three depths to the net: the nodes on the left (depth 1) receive message traffic from external sources; the middle node (depth 2) receives traffic only from nodes at depth 1; and the nodes on the right (depth 3) receive traffic only from the central node at depth 2. The quantity of interest in all cases is the distribution (or histogram) of total time spent by messages in the central node. The nodes at depth 3 serve as destinations for all messages.[1] In all runs, the total capacity of the net was held fixed and was broken into two equal parts, each of C bits/sec. The capacity was assigned so that the capacities of all channels connecting depths 1 and 2 totaled C, and similarly for the sum for channels connecting depths 2 and 3. When more than one channel connected adjacent depths, the capacity C was split equally among these channels. These comments apply also to configurations a and b, except that the node at depth 2 is omitted.

Figure 3.6 shows the results of the simulation. Figure 3.6a and b shows the histogram of message delay in passing through the first node for the net in Fig. 3.5a and b. All other parts of Fig. 3.6 show the histogram of message delay in passing through the single node at depth 2. In all cases, the quantity $\rho = \gamma/\mu C$ is displayed on the histogram itself, where ρ pertains to a single channel emanating from the black node.

Note that Fig. 3.6a and b is essentially exponential distributions as, of course, it should be [see Eq. (A.5)]. Figure 3.6c exposes the behavior of $p(R)$, the distribution whose analytic form we were not able to obtain. We note that as ρ increases there is a marked difference in behavior between $p(Q)$ in Fig. 3.6a and $p(R)$ in Fig. 3.6c. Supplying two input channels (Fig. 3.6d) and five input channels (Fig. 3.6e) changes the nature of the difference between these figures and Fig. 3.6a; however, this difference is considerable even at moderate values of ρ.

On the other hand, the introduction of even two paths (Fig. 3.6f) out of the central node at depth 2 results in a tremendous reduction in the difference between the behaviors of the first and second nodes. Adding five exits (Fig. 3.6g) from this node increases the similarity even further. In fact, as this point, one is hard pressed to distinguish between Fig. 3.6a, b, g. Figure 3.6h shows, for completeness, the five-input five-output case, which behaves very much in the same way as the one-input five-output case shown in Fig. 3.6g. The horizontal distance

---

[1] That is, equal traffic rates are applied to nodes at depth 1, and each node at depth 3 serves as a destination, all receiving the same average number of messages.

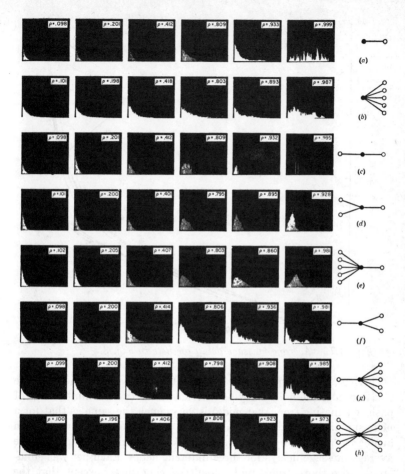

Fig. 3.6. *Histograms of message delay from digital simulation.*

between bars on each histogram indicates the scale expansion used in displaying the histogram; that is, the spacing between adjacent bars represents one unit of delay.

In Fig. 3.7 we plot the average message delay for these configurations as obtained from the simulation.[1] This figure gives quantitive reference to the comments made above. In particular, we note the essential similarity of the average message delays for configurations *a, b, f, g,* and *h.*

[1] The quantity C in this figure is taken to be the capacity of a single channel entering a node at depth 3.

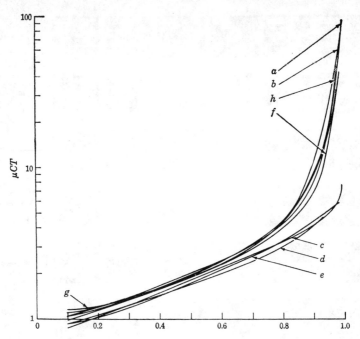

Fig. 3.7.   *Average message delay for nets in Fig. 3.5*

Having shown experimentally that the behavior of the message delay for a single node carrying internal traffic (with a multiplicity of paths emanating from that node) is very much the same as that for a node supplied exclusively with external traffic, we must now show an analogous result for the entire net.   We proceed by comparing the average message delays for three different nets.   Each net was simulated twice, both times under identical conditions, except that the independence assumption was made in only one of the two cases.   The detailed description of each net is given in Sec. 7.1; for our present purposes these details are not of importance.   Figure 3.8 shows[1] the effect of introducing the independence assumption for a particular traffic matrix τ_1 (which represented a rather nonuniform traffic); Fig. 3.9 shows a similar graph for the uniform traffic matrix τ_3 (once again, see Sec. 7.1 for a full description of these nets and traffic matrices). The important observation to make is that, in all cases, the introduc-

---

[1] T is the message delay averaged over all origin-destination pairs, $1/\mu$ is the average message length in bits, C is the total channel capacity assigned to the net, and γ is the total number of messages per second entering the net from external sources.

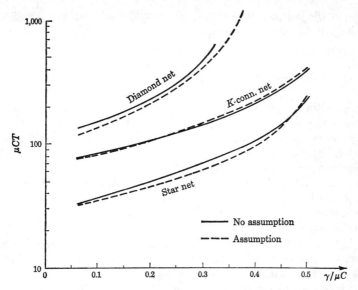

Fig. 3.8. *Effect of independence assumption (nonuniform traffic).*

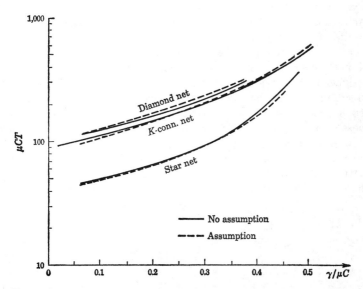

Fig. 3.9. *Effect of independence assumption (uniform traffic).*

tion of the independence assumption resulted in a rather insignificant change in the average message delay.

3.6 Summary

An attempt has been made in this chapter to show the essential complexity and intractability of a direct mathematical analysis of the general communication net. Even the simpler class of tandem nets has led to rather limited analytical results. The source of difficulty has been shown to be the dependence between the interarrival times and the lengths of internal message traffic. We have found, however, that the introduction of the independence assumption removes this dependency and produces a model which behaves essentially the same as the original model with respect to average message delay.

It is now appropriate to discuss the mathematical model which results from the introduction of the independence assumption. Specifically, if we consider only fixed routing procedures,[1] we recognize that the internal traffic flowing into each channel is statistically equivalent to the external traffic entering the net; i.e., the interarrival times and message lengths are independent and are chosen from exponential distributions. Thus, each channel satisfies the conditions of the single exponential channel described in Appendix A. This observation, coupled with Theorem A.1 (due to Burke), allows us to consider each channel (or node) separately in the mathematical analysis; the results of this individual analysis, of course, must yield the equations described in Appendix A (i.e., for the single exponential channel). We cannot, however, stop here and assume that we have answered the designer's questions as posed in Sec. 1.4. We must consider the effects of various channel capacity assignments, routing procedures, topological structures, and priority disciplines as well. Fortunately, these considerations are vastly simplified by the use of the independence assumption. We note at this point that the introduction of various alternate-routing procedures may easily complicate the mathematics once again; we therefore find that in many cases we still rely on the simulation procedure for results.

[1] See the footnote on page 21.

chapter 4

Some New Results for Multiple-channel Systems

We recognize that the problems associated with a multiterminal communication net appear to be too complex for analysis in an exact mathematical form. That is, the calculation of such quantities as the multivariate distribution of traffic flow through a large (or even small) network is extremely difficult.[1] However, the introduction of the independence assumption into our model simplifies matters considerably. Specifically, we are now able to carry out an analysis of message delay on a node-by-node basis, as discussed in Sec. 3.6. It would be appropriate, at this point, for the reader to acquaint himself with the simple results from queueing theory which are presented in Appendix A.

In the present chapter, we derive some new results for simple multiple-channel systems under a first-come first-served discipline and consider optimum channel capacity assignments. As a preface, we briefly state the problems considered and the solutions obtained herein. Specifically, we present a canonical representation for the utilization factor in a single-node system which has N output channels, each of arbitrary channel capacity. We then proceed to determine that number N of output channels from a single node which minimizes the time that a message spends in the node, subject to the constraint that each channel is assigned a capacity C/N. A set of trading relations

---

[1] As discussed at length in Chap. 3.

among message delay, channel capacity, and total traffic handled is developed next from some well-known equations. A result is then obtained which gives that assignment of channel capacities to a set of N independent single-output-channel nodes which minimizes the message delay averaged over the set of N nodes, subject to the constraint that the sum of the assigned channel capacities is constant. The optimum assignment of the traffic pattern is discussed under some interesting constraints.   We then consider the more general case of an interconnected net subject to a fixed routing procedure[1] and find that the optimum channel capacity assignment is almost the same as for the unconnected net; the expressions now involve a new quantity $\bar{n}$, the average path length for messages.   Finally, we generalize the cost function applied to the previous results and conclude with a theorem which describes the optimum channel capacity assignment for this case.

4.1   A Canonical Representation for the Utilization Factor

Whereas it is well known that, for a single-channel system, the utilization factor ρ represents the fraction of time that the channel is in use, there has not been a similar representation available for a general multiple-channel system.[2]   One suspects that there should be an extended interpretation for such a system; indeed, there is, as is stated in the following:

Theorem 4.1[3]

Consider an N-channel service facility of total capacity C bits/sec (the distribution of the total capacity C among the N channels being completely arbitrary) with Poisson arrivals at an average arrival rate of λ messages per second, message lengths distributed exponentially with mean length $1/\mu$ bits, and an arbitrary queue discipline (with the restrictions that there be no defections from the system and, if preemption is allowed, it must be preemptive

---

[1] See the footnote on page 21.

[2] For the special multiple-channel case wherein all channels have identical capacities (as in Sec. 4.2), it is well known (see, for example, Morse [20, p. 102]) that ρ is the average fraction of busy channels. We are giving a more general result which allows an arbitrary distribution of capacity among the channels.

[3] Appendix B contains the proof of this theorem.

resume[1]). Define, as usual, the utilization factor

$$\rho = \frac{\lambda}{\mu C}$$

Then
$$\rho = 1 - \sum_{n=0}^{\infty} \frac{\bar{C}_n}{C} P_n \qquad (4.1)$$

provided
$$\rho < 1$$

where $Q_n(x) = \mathrm{P_r}[\text{sum of capacities of all unused channels is less than } x, \text{ given } n \text{ messages in system}]$

$\bar{C}_n = \int_0^C x \, dQ_n(x) = \text{expected value of unused capacity given } n \text{ messages in system}$

$P_n = \mathrm{P_r}[\text{finding } n \text{ messages in system in steady state}]$

Essentially, this theorem states that[2]

$$\rho = E \text{ (used normalized capacity)}$$

where the normalization is with respect to the total capacity C. This theorem not only gives one a physical interpretation of the utilization factor for a multiple-channel system, but also gives a quite useful alternative analytic expression for the utilization factor. The proofs of certain theorems in Chap. 6 depend upon this representation.

4.2 Optimum Number of Channels for a Single-node Facility

Consider a pair of nodes in a large communication net, the first transmitting a message destined for the second. One can inquire as to the appearance of the rest of the net from the point of view of the transmitting node. It does not seem unreasonable to consider that the rest of the net offers, to the message, a number N of "equivalent" alternative paths from the first node to the second; the equivalence is a very gross simplification of the actual situation, which, nevertheless, serves a useful purpose. Thus, the node under consideration reduces itself to a multiple-channel system which we now proceed to discuss. Assume that we have N channels emanating from this node, each of capacity C/N bits/sec, with Poisson arrivals at an average arrival rate of λ messages per second, and with all message lengths exponentially distributed with mean length $1/\mu$ bits. The queue discipline is taken to be first-come first-served, wherein the message at the head of the queue accepts the first channel to become available. If a

---

[1] See Sec. 5.1 for a precise definition of these terms.

[2] The notation $E(x)$ is to be interpreted as "expected value of x."

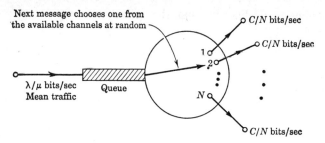

Fig. 4.1. *N-channel node considered in Theorem 4.2.*

message enters the system when more than one channel is free, it chooses one from this set according to a uniform distribution.[1] Such an arrangement is shown in Fig. 4.1.

For given values of λ, μ, and the total capacity C of the node, the question as to the proper choice for N (the total number of channels) presents itself.[2] Specifically, let us inquire as to the value of N which minimizes T, the mean total time spent in the node (i.e., time spent waiting for a free channel plus time spent in transmission over that channel). Again we define $\rho = \lambda/\mu C$. Appendix A presents the solution for T as well as a number of other pertinent expressions. We are now ready to state:

Theorem 4.2[3]

The value of N which minimizes T for all $0 \leq \rho < 1$ is

$$N = 1$$

A facility with more than one channel is nonoptimum in the sense described above because the efficiency of a node is related to its transmitting rate; if we have only one channel, we must transmit at a rate of C bits/sec whenever there are any messages in the system; if we have N channels ($N > 1$), there will be situations in which less than N channels are occupied, and we shall then be transmitting at a rate less than C.

This result says, in essence, that whenever possible one should design a multiple-channel system (whose total capacity is fixed) with

[1] Recall that we are assuming that all channels leading out of the node go to "equivalent" destinations, and so a message is willing to accept any channel at all.

[2] Morse [20, p. 103] discusses this problem.

[3] See Appendix B for proof of this theorem.

as few channels as the physical constraints of the network allow (the limiting case of one channel is, as stated above, optimum).

4.3 Trading Relations among Rate, Capacity, and Message Delay

A consideration of the trading relations among the number of messages handled, the expected delay experienced by these messages, and the channel capacity of the facility will now be undertaken. Let us consider two different single exponential channel facilities, as shown in Fig. 4.2. We also consider the two quantities T_i and W_i, where, once again,

$T_i = E$ (total time that a message spends in passing through node i)

and where we define

$W_i = E$ (time that a message spends on queue in node i)

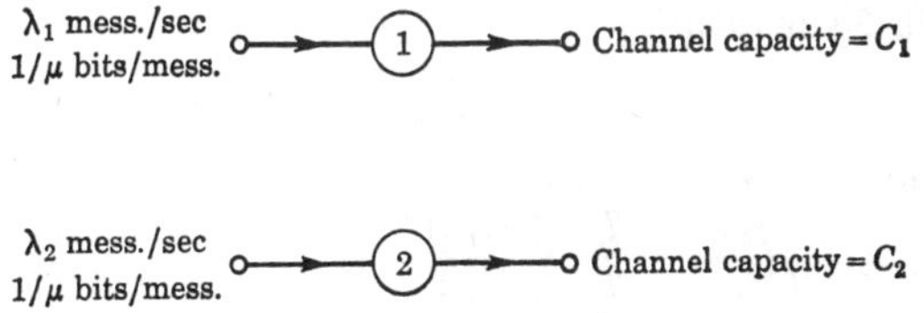

λ_1 mess./sec
$1/\mu$ bits/mess. o———(1)———o Channel capacity $= C_1$

λ_2 mess./sec
$1/\mu$ bits/mess. o———(2)———o Channel capacity $= C_2$

Fig. 4.2. *Two single exponential channel facilities.*

We assume that the messages arriving at both nodes have the same average length ($1/\mu$ bits per message) but different Poisson arrival rates (λ_i messages per second), where $i = 1, 2$.

What is of interest to us is the relative behavior of these two systems with regard to their message rate, message delay, and channel capacity. Specifically, we desire quantitative relations for λ_2/λ_1, T_2/T_1, W_2/W_1, and C_2/C_1. By straightforward use of Eq. (A.6), we find that

$$\frac{T_2}{T_1} = \frac{C_1}{C_2} \frac{1 - \rho}{1 - (\lambda_2 C_1/\lambda_1 C_2)\rho}$$

and

$$\frac{W_2}{W_1} = \frac{\lambda_2 C_1{}^2}{\lambda_1 C_2{}^2} \frac{1 - \rho}{1 - (\lambda_2 C_1/\lambda_1 C_2)\rho}$$

where

$$\rho = \frac{\lambda_1}{\mu C_1}$$

and

$$W_i = T_i - \frac{1}{\mu C_i}$$

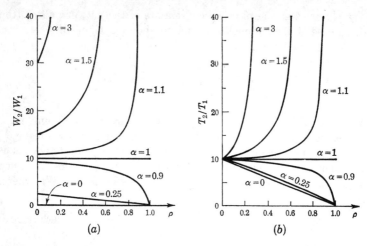

Fig. 4.3. *Trading curves among message delay, channel capacity, and total traffic handled.*

If we make the substitution

$$\alpha = \frac{\lambda_2 C_1}{\lambda_1 C_2}$$

we then obtain

$$\frac{T_2}{T_1} = \frac{C_1}{C_2} \frac{1 - \rho}{1 - \alpha\rho} \qquad (4.2)$$

and

$$\frac{W_2}{W_1} = \alpha \frac{C_1}{C_2} \frac{1 - \rho}{1 - \alpha\rho} \qquad (4.3)$$

The general behavior of these relations is shown in Fig. 4.3, where we have taken $C_1/C_2 = 10$ for purposes of illustration; this same figure may be used for any $C_1/C_2 = M$ by multiplying the vertical scale by $C_1/10C_2$. Equations (4.2) and (4.3) give the desired trading relations. Thus, the effect of varying one of the parameters may be seen quite simply from these curves.

An important observation may now be made. Let us assume that C_1 is M times as large as C_2. Now, one might imagine that reducing the input rate of messages to the second (lower-capacity) system by the same ratio (that is, $\lambda_2 = \lambda_1/M$) would leave the relative message delay constant. However, it is clear from Eqs. (4.2) and (4.3) that this is not so. In particular, in this case, we see that

$$\frac{T_2}{T_1} = M$$

and

$$\frac{W_2}{W_1} = M$$

This result, somewhat surprising at first, has certain implications about the design of any queueing facility. In particular, it states that large facilities with large arrival rates of units to be serviced perform better than small facilities with proportionally smaller arrival rates of units.[1] The increase in performance in this case is equal to the ratio of the capacities of the two facilities and is independent of the value of ρ.

It is also clear from the curves in Fig. 4.3 that there exists a range for ρ in which it is possible to reduce the input message rate sufficiently so that a reduction in capacity may be made with no increase in message delay (that is, $T_2 = T_1$), namely, $1 - C_2/C_1 \leq \rho < 1$.

4.4 Optimum Assignment of Channel Capacity

We first consider a situation in which there are N separate single exponential channel facilities. The ith node has a Poisson arrival rate of λ_i messages per second, each message having an exponentially distributed length of mean $1/\mu_i$ bits; the channel capacity associated with the ith node is C_i. All nodes behave independently of each other; however, they are mutually coupled by the following linear constraint on their capacities:

$$C = \sum_{i=1}^{N} C_i \tag{4.4}$$

That is, there is distributed throughout the N channels a total capacity of C bits/sec. The system under consideration is shown in Fig. 4.4.

[1] Feller [18, p. 420] discusses a comparison of this type.

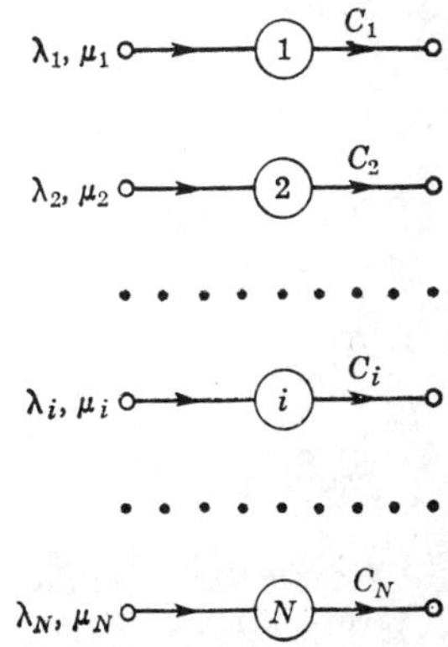

Fig. 4.4. *System of N separate single-channel facilities.*

For any assignment of the C_i which satisfies Eq. (4.4), there is defined

$$T_i = E \text{ (total time that a message spends in passing through node } i)$$

One may ask about that particular assignment of the C_i which satisfies Eq. (4.4) and which also minimizes the average (over the index i) of the set of numbers T_i. Specifically, we define this average to be

$$T = \sum_{i=1}^{N} \frac{\lambda_i}{\lambda} T_i \tag{4.5}$$

where[1]
$$\lambda = \sum_{i=1}^{N} \lambda_i \tag{4.6}$$

Note that the weighting factor λ_i/λ for T_i has been chosen in the obvious way to be proportional to the number of messages which pass through node i. The solution to this problem is stated in:

Theorem 4.3[2]

The assignment of the set C_i which minimizes T and which satisfies Eq. (4.4) is

$$C_i = \frac{\lambda_i}{\mu_i} + C(1 - \rho) \frac{\sqrt{\lambda_i/\mu_i}}{\sum_{j=1}^{N} \sqrt{\lambda_j/\mu_j}} \tag{4.7}$$

provided that

$$C > \sum_{i=1}^{N} \frac{\lambda_i}{\mu_i} \tag{4.8}$$

where
$$\rho = \frac{\lambda}{\mu C} \tag{4.9}$$

and
$$\frac{1}{\mu} = \sum_{i=1}^{N} \frac{\lambda_i}{\lambda} \frac{1}{\mu_i} \tag{4.10}$$

With this optimum assignment, we find that

$$T_i = \frac{\sum_{j=1}^{N} \sqrt{\lambda_j/\mu_j}}{C(1 - \rho) \sqrt{\lambda_i \mu_i}} \tag{4.11}$$

and
$$T = \frac{\left(\sum_{i=1}^{N} \sqrt{\lambda_i/\mu_i} \right)^2}{\lambda C(1 - \rho)} \tag{4.12}$$

---

[1] Note that $\lambda = \gamma$ in this special configuration (see definitions in Sec. 1.7).
[2] See Appendix B for proof of this theorem.

We note that the optimum assignment operates in the following way. Each channel is first apportioned just enough capacity to satisfy its average required flow of λ_i/μ_i bits/sec. After this apportionment, there remains an excess capacity $C - \sum_{i=1}^{N} \lambda_i/\mu_i = C(1 - \rho)$ which is then distributed among the channels in proportion to the square roots of their average flows λ_i/μ_i. Equation (4.8) expresses the obvious condition that there be enough capacity initially to satisfy the minimum requirements of the average flow in each node.

Having obtained the optimum assignment for C_i, we now inquire as to the optimum distribution of the λ_i. We consider the case in which $\mu_i = \mu$ for $i = 1, 2, \ldots, N$. Let us assume that we have some freedom in distributing the λ_i among the N nodes, subject to the constraint expressed in Eq. (4.6) and the additional constraints that

$$\lambda_i \geq k_i \qquad i = 1, 2, \ldots, N \qquad (4.13)$$

where, for convenience, we order the subscript i such that

$$k_1 \geq k_2 \geq \cdots \geq k_N \geq 0 \qquad (4.14)$$

and where, obviously,

$$\sum_{i=1}^{N} k_i \leq \sum_{i=1}^{N} \lambda_i = \lambda$$

The set of numbers k_i represents lower bounds on the traffic flow into any node; this set of constraints corresponds to a sensible physical limitation on traffic flow. The solution to this problem is stated in:

Theorem 4.4[1]

The distribution of λ_i which minimizes T in Eq. (4.12) subject to the constraints expressed by Eqs. (4.6) and (4.13) is, for $\mu_i = \mu$,

$$\lambda_i = \begin{cases} \lambda - \sum_{j=2}^{N} k_j & i = 1 \\ k_i & i = 2, 3, \ldots, N \end{cases} \qquad (4.15)$$

Now, for all $k_i = 0$ [i.e., no constraint due to Eq. (4.13)], we find that all the traffic should be assigned to (any) one of the channels, and this channel should be assigned the total capacity C. For the general case as expressed by Theorem 4.4, we see that after the constraint due to Eq. (4.13) is satisfied in the minimum sense (that is, $\lambda_i = k_i$), all the additional traffic should be assigned to that channel which has the largest k_i (namely, channel C_1).

[1] See Appendix B for proof of this theorem.

We now consider the more general case of an interconnected net (as, for example, in Fig. 1.3) with N channels subject to a fixed routing procedure.[1] Since we accept the independence assumption, all message lengths are chosen independently at each node from an exponential distribution. Furthermore, the externally applied traffic is Poisson in nature. Consequently, Theorem A.1 (due to Burke) is satisfied, and we find that the interarrival times for message arrivals throughout the net are also Poisson. This being the case, the optimum channel capacity assignment for the net, with a fixed total capacity C, is described by an equation similar to Eq. (4.7).[2] The interpretation of λ_i is, as before, the average arrival rate of messages to the ith channel; further, we take $\mu_i = \mu$ for all i. The average message delay T now must be carefully defined as

$$ T = \sum_{j,k} \frac{\gamma_{jk}}{\gamma} Z_{jk} \qquad (4.16) $$

where γ_{jk} = average number of messages entering network, per second, with origin j and destination k

$\gamma = \sum_{j,k} \gamma_{jk}$

Z_{jk} = average message delay for messages with origin j and destination k

That is, T is appropriately defined as the overall average message delay, where the weighting factor for Z_{jk} is taken to be proportional to the number of messages which must suffer the delay Z_{jk}. For any pair jk, the quantity Z_{jk} is composed of the sum of the average delays encountered in passing through each channel on the fixed route from node j to node k. If we break Z_{jk} into such components, and if we also form T by summing over the individual delays suffered at each channel in the net (instead of summing the delays for origin-destination pairs), we immediately see that

$$ T = \sum_i \frac{\lambda_i}{\gamma} T_i \qquad (4.17) $$

where clearly λ_i is the sum of all γ_{jk} for which the (fixed) jk route includes channel i. Thus we note that T is defined in a consistent manner [that is, $\lambda = \gamma$ for the net in Fig. 4.4, and so Eqs. (4.5) and (4.17) are equivalent]. We may now state one of our fundamental results:

---

[1] See the footnote on page 21.
[2] See Theorem 4.5 below.

Theorem 4.5[1]

For a net as described above, with a fixed routing procedure, the optimum channel capacity assignment is

$$C_i = \frac{\lambda_i}{\mu} + C(1 - \bar{n}\rho) \frac{\sqrt{\lambda_i}}{\sum\limits_{j=1}^{N} \sqrt{\lambda_j}}$$

With this optimum assignment,

$$T_i = \frac{\sum\limits_{j=1}^{N} \sqrt{\lambda_j}}{\mu C(1 - \bar{n}\rho) \sqrt{\lambda_i}}$$

and the average message delay T is

$$T = \frac{\bar{n} \left(\sum\limits_{i=1}^{N} \sqrt{\lambda_i/\lambda} \right)^2}{\mu C(1 - \bar{n}\rho)} \tag{4.18}$$

where $\bar{n} = \lambda/\gamma$ is the *average path length*[2] for messages.

This theorem shows the strong dependence of T on the average path length $\bar{n}$, as well as on the distribution of λ_i. The significance of this result is discussed in Chap. 7 in conjunction with the results of the simulation experiments.

4.5   Conclusions and Extensions

A number of different questions have been posed in this chapter, and we now summarize some of the conclusions we have drawn.

If we consider the results expressed by Theorems 4.2 and 4.4 and Eq. (4.2), we may draw a unifying conclusion: these results all indicate that delay is minimized in a queueing process when traffic is concentrated into as few channels as is physically possible! The underlying constraint which forced this result to the surface is that expressed in Eq. (4.4), which insists on a constant total channel capacity assigned to the system of nodes. This conclusion is verified by the simulation results presented in Chap. 7.

Furthermore, we have solved for the optimum[3] channel capacity

---

[1] See Appendix B for proof of this theorem.
[2] In Appendix B we carefully define $\bar{n}$ and prove that it is equal to λ/γ.
[3] Optimum in the sense of minimizing the average message delay T.

assignment for a communication net with fixed routing subject to the constraint of fixed total channel capacity.

An extremely useful extension to the results of Sec. 4.4 will now be described. Specifically, we have introduced the constraint, expressed by Eq. (4.4), which limits the total channel capacity, in bits, that may be assigned to the system. The rationale for this constraint is that the total assigned capacity is one measure of the cost of constructing the system. This function assumes no measure of the cost per unit capacity associated with each channel. Indeed, a more realistic cost function would be one which included, as a factor, some function d_i of the ith channel. In particular, we now offer, as an alternative to the constraint expressed in Eq. (4.4), the following condition:

$$D = \sum_{i=1}^{N} d_i C_i \tag{4.19}$$

where C_i is the channel capacity of the ith transmission channel, and d is a function independent of the capacity C_i which reflects the cost, say in dollars, of supplying one unit of channel capacity to the ith channel. The quantity D represents the total number of dollars that is available to spend in supplying the N-channel system with the set of capacities C_i ($i = 1, 2, \ldots, N$). In this case, we develop a theorem, analogous to Theorem 4.5, as follows:

Theorem 4.6[1]

The assignment of the set of channel capacities C_i to a communication net with a fixed routing procedure (such as is described for Theorem 4.5) which minimizes T [see Eq. (4.17)] subject to the constraint expressed in Eq. (4.19) is

$$C_i = \frac{\lambda_i}{\mu} + \left(\frac{D_e}{d_i}\right) \frac{\sqrt{\lambda_i d_i}}{\sum\limits_{j=1}^{N} \sqrt{\lambda_j d_j}} \tag{4.20}$$

With this optimum assignment,

$$T_i = \frac{\sum\limits_{j=1}^{N} \sqrt{\lambda_j d_j}}{\mu D_e \sqrt{\lambda_i/d_i}} \tag{4.21}$$

and

$$T = \frac{\bar{n} \left(\sum\limits_{i=1}^{N} \sqrt{\lambda_i d_i/\lambda}\right)^2}{\mu D_e} \tag{4.22}$$

[1] See Appendix B for proof of this theorem.

provided $$D_e > 0$$

and where $$D_e = D - \sum_{j=1}^{N} \frac{\lambda_j d_j}{\mu} \qquad (4.23)$$

The analogy among Theorems 4.3, 4.5, and 4.6 is clear.[1] Some interesting special cases of d_i are listed below, where we define m_i as the length of the ith channel.

1. $d_i = m_i$. This puts cost proportional to length times capacity, such as is the case in the laying of telephone cables, wherein one of the major costs is the copper cost.
2. $d_i = m_i^4$. This puts cost proportional to the fourth power of the length times the capacity, which approximates the situation in an *Echo*-type passive satellite.

Obviously, the actual form of the set of cost functions d_i depends upon the particular communication system involved. The implications of this new constraint [Eq. (4.19)] bear further investigation for future research.

[1] For example, Theorem 4.5 is the special case of Theorem 4.6 wherein $D = C$ and $d_i = 1$ for all i.

chapter **5**

Waiting Times
for Certain
Queue Disciplines

We now explore the manner in which message delay is affected when one introduces a priority structure (or queue discipline) into the set of messages in a single-node facility with a single transmission (or service) channel. In this chapter we shall present some newly derived results for certain queue disciplines; some previously published results are also included for completeness.

In communication nets such as we are considering, messages are forced to form a queue while awaiting passage through a transmission facility, and often a priority discipline describes the queue structure. The rule for choosing which message to transmit next is frequently based on a priority system similar to those studied in this chapter. Generally, one breaks the message set into P separate groups, the pth group ($p = 2, 3, \ldots, P$) being given preferential treatment over the $(p - 1)$st group, etc. Introducing a priority structure into the message set influences the expected value of the time that each priority group spends in the queue. It is this statistic which is of interest to us and which will be solved for. An understanding of the effects of a priority discipline at the single-node level is necessary before one can make any intelligent statements about the multinode case.

A new result for a delay-dependent priority system is described in which a message's priority is increased, from zero, linearly with time in proportion to a rate assigned to the message's priority group. The usefulness of this new priority structure is that it provides a number of degrees of freedom with which to manipulate the relative waiting times for each priority group.

An interesting new law of conservation is also proven which constrains the allowed variation in the average waiting times for any one of a wide class of priority structures. As a result of this law, a number of general statements can be made regarding the average waiting times for any priority structure which falls in this class. A system with a time-shared service facility is also investigated. This system results in shorter waiting times for "short" messages and longer waiting times for "long" messages; interestingly enough, the critical message length which distinguishes short from long turns out to be the average message length for the case of geometrically distributed lengths.

It is assumed throughout that the systems under consideration are in steady-state equilibrium. In general, this is equivalent to requiring that the system has been operating for a long time and that $\rho < 1$, where ρ, once again, is the product of the average arrival rate of messages and the expected transmission time for each message. However, in some of the priority systems studied, it is possible to have $\rho \geq 1$ and still obtain a steady-state type solution for some of the higher-priority messages. For a full discussion of this aspect of the problem, the reader is referred to Phipps [47].

5.1 Priority Queueing

Priority queueing refers to those disciplines in which an entering message is assigned a set of parameters (either at random or based on some property of the message) which determine its relative position in the queue. This position will vary as a function of time, owing to the appearance of messages of higher priority in the queue. At any time t, the priority of a particular message is calculated as a function of the assigned parameters; the higher the value obtained by this function, the higher the priority. That is, the notation used is such that a message with priority q_2 is given preferential treatment over a message with priority q_1, where $q_2 > q_1$. In the fixed-priority system to be discussed presently, this means that a message from the pth priority group has a higher priority than a message from the $(p-1)$st group. A tie for highest priority is broken by a first-come first-served doctrine.

Let there be a total of P different priority classes. Messages from priority class p ($p = 1, 2, \ldots, P$) arrive in a Poisson stream at rate λ_p messages per second; each message from priority class p has a total required processing time[1] selected independently from an exponential distribution with mean $1/\mu_p$. We define

$$\lambda = \sum_{p=1}^{P} \lambda_p \tag{5.1}$$

$$\frac{1}{\mu} = \sum_{p=1}^{P} \frac{\lambda_p}{\lambda} \frac{1}{\mu_p} \tag{5.2}$$

$$\rho_p = \frac{\lambda_p}{\mu_p} \tag{5.3}$$

$$\rho = \frac{\lambda}{\mu} = \sum_{p=1}^{P} \rho_p \tag{5.4}$$

$$W_0 = \sum_{p=1}^{P} \frac{\rho_p}{\mu_p} \tag{5.5}$$

We further define

W_p = expected value of time spent in queue for a message with assigned parameter p

W_0 may be interpreted as the expected time required to complete service on the message found in service upon entry of a new message to the system.

We consider four types of priority systems. In two of the systems we assume that once a message enters the processing stage, it cannot be interrupted, and the entire processing effort is devoted to completing its transmission. This rule defines a system with no *preemption*. In contrast, two types of queueing systems studied do allow preemption; i.e., a message will be taken out of the processing (or transmission) stage immediately when another message of higher priority appears in the queue. Since we assume that each message has a fixed servicing time (chosen from some exponential distribution) associated with it, we must further assume that when a preempted message reenters the service facility, its servicing is started at the point at which it was

---

[1] In the application of interest, wherein messages are passing through a transmission facility, the average processing time (time spent in the transmission channel) is $1/\mu_p C$, where, once again, C is the capacity of the channel in bits/sec, and where $1/\mu_p$ is the average message length in bits. However, for the purposes of this chapter, it is convenient to suppress the parameter C, and so we assume that $C = 1$ throughout (with no loss of generality). If one wishes to reintroduce it, one need merely multiply every μ_p by C.

interrupted when preemption occurred (this is referred to as *pre-emptive resume*).[1]

The other distinguishing feature among the systems is the form of the priority assignment. In two of the systems the priority assignment for any message remains fixed in time; i.e., an entering message is assigned a number (say p) which is to be the fixed value of its priority. We refer to such systems as *fixed-priority* systems. In the other two systems, the priority assignment varies linearly with time. In particular, a message entering the queue at time T is assigned a number b_p, where $0 \leq b_1 \leq b_2 \leq \cdots \leq b_P$, and the priority $q_p(t)$ at time t associated with that message is calculated as follows:

$$q_p(t) = (t - T)b_p \tag{5.6}$$

where t ranges from T up to the time at which this message's service is completed. This system is referred to as a *delay-dependent priority* system.

Thus, in summary, the four priority systems considered are:

1. Fixed-priority system with no preemption
2. Fixed-priority system with preemption
3. Delay-dependent priority system with no preemption
4. Delay-dependent priority system with preemption

In all four systems, when a new message is to enter the processing (servicing) facility, the highest-priority message is chosen.

We first consider the fixed-priority system with no preemption. The results presented here are due to Cobham [48]; his notation has been altered slightly to correspond to our convention of ordering the priorities.[2]

Theorem (*due to Cobham*)

For $0 \leq \rho$,

$$W_p = \begin{cases} \dfrac{f\rho_{j-1}/\mu_{j-1} + \sum\limits_{i=j}^{P} \rho_i/\mu_i}{\left(1 - \sum\limits_{i=p+1}^{P} \rho_i\right)\left(1 - \sum\limits_{i=p}^{P} \rho_i\right)} & \text{for } p \geq j \\[4mm] \infty & \text{for } p < j \end{cases} \tag{5.7}$$

[1] Such a procedure requires some additional bookkeeping at the transmission and receiving facilities to keep track of the individual messages. We do not consider the problems associated with this bookkeeping.

[2] Phipps [47] has used Cobham's results as a basis for the treatment of a particular variety of machine-repair problem in which shortest jobs receive highest priority. His priority ordering passes, therefore, from a discrete set to a continuum.

where j = smallest positive integer such that $\sum\limits_{i=j}^{P} \rho_i < 1$

and
$$f = \begin{cases} 0 & \rho < 1 \\ \dfrac{1 - \sum\limits_{i=j}^{P} \rho_i}{\rho_{j-1}} & \rho \geq 1 \end{cases} \tag{5.8}$$

Note that for $\rho < 1$ the numerator of W_p becomes merely W_0. A graph of the family W_p is plotted in Fig. 5.2a to d for a particular set of parameters.

We now consider the fixed-priority system with preemption. The results presented here were derived independently by the author and correspond to the preemptive-resume case considered by White and Christie [49] with exponential service times.[1] We define

$$W_p = T_p - \frac{1}{\mu_p}$$

where T_p is the expected value of the total time spent in the system by a message of priority p. Two forms are given for W_p, the first recursively in terms of the W_i for the higher-priority messages, and the second recursively in terms of the W_i for the lower-priority messages.

Theorem 5.1[2]

For a fixed-priority system with preemption and $0 \leq \rho$

$$W_p = \begin{cases} \dfrac{\dfrac{\rho_p}{\mu_p} + \sum\limits_{i=p+1}^{P} \rho_i\left(\dfrac{1}{\mu_p} + \dfrac{1}{\mu_i}\right) + \sum\limits_{i=p+1}^{P} \rho_i W_i}{1 - \sum\limits_{i=p}^{P} \rho_i} & p \geq j \\ \\ \infty & p < j \end{cases} \tag{5.9}$$

or

$$W_p = \begin{cases} \dfrac{\dfrac{s_j}{1 - s_j} \sum\limits_{i=j}^{P} \dfrac{\rho_i}{\mu_i} + \dfrac{\rho_p}{\mu_p} + \sum\limits_{i=p+1}^{P} \rho_i\left(\dfrac{1}{\mu_p} + \dfrac{1}{\mu_i}\right) - \sum\limits_{i=j}^{p-1} \rho_i W_i}{1 - \sum\limits_{i=p+1}^{P} \rho_i} & p \geq j \\ \\ \infty & p < j \end{cases} \tag{5.10}$$

[1] The form for W_p in Eq. (5.9) was given by White and Christie. The form in Eq. (5.10) is new.

[2] See Appendix C for proof of this theorem.

where j is as defined above in Cobham's results, and

$$s_j = \sum_{i=j}^{P} \rho_i \qquad (5.11)$$

Note that for $\rho < 1$ we obtain $j = 1$, $s_j = s_1 = \rho$, and $\sum_{i=j=1}^{P} \rho_i/\mu_i = W_0$. A graph of the family W_p is plotted in Fig. 5.3a to d for the same parameters as Fig. 5.2.

We next consider the delay-dependent priority system. As defined, a message from the pth priority group entering the queue at time T is assigned a number b_p, where $0 \leq b_1 \leq b_2 \leq \cdots \leq b_P$; the priority $q_p(t)$ associated with that message at time t is calculated from

$$q_p(t) = (t - T)b_p$$

where t ranges from T until the time at which this message's service is completed. Figure 5.1 shows the manner in which this priority structure allows interaction between the priority functions for two messages. Specifically, at time T, a message from priority group p_1 arrives and attains priority at a rate equal to $(t - T)b_{p_1}$. At time T', a different message enters from a higher-priority group p_2; that is, $p_2 > p_1$. When the service facility becomes free, it chooses for transmission that message in the queue with the highest instantaneous priority. Thus, in our example, the first message will be chosen in preference to the second if the transmission facility becomes free at any time between T and T_0; at any time after T_0, the second message will be chosen in preference to the first.

For the delay-dependent priority system without preemption, we give two derived forms for W_p; one is a recursive form in terms of the

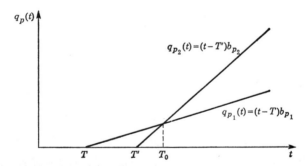

Fig. 5.1. *Interaction between priority functions for the delay-dependent priority system.*

W_i for the lower-priority messages, while the other, more complicated, expression is the solution of the recursive equations.

Theorem 5.2[1]

For the delay-dependent priority system with no preemption and $0 \le \rho < 1$,

$$W_p = \frac{W_0/(1-\rho) - \sum_{i=1}^{p-1} \rho_i W_i (1 - b_i/b_p)}{1 - \sum_{i=p+1}^{P} \rho_i (1 - b_p/b_i)} \tag{5.12}$$

or $\quad W_p = \dfrac{W_0}{1-\rho} \dfrac{1}{D_p} \left[1 + \right.$

$$\left. \sum_{j=1}^{p-1} \sum_{0 < i_1 < i_2 < \cdots < i_j < p} F_{i_1}(i_2) F_{i_2}(i_3) \cdots F_{i_j}(p) \right] \tag{5.13}$$

where $\qquad D_p = 1 - \sum_{i=p+1}^{P} \rho_i \left(1 - \dfrac{b_p}{b_i} \right) \tag{5.14}$

and $\qquad F_k(n) = - \dfrac{\rho_k}{D_k} \left(1 - \dfrac{b_k}{b_n} \right) \tag{5.15}$

A graph of the family W_p is plotted in Fig. 5.4*a* and *b*. It is interesting to note the extremely simple dependence of W_p on the parameters b_i (namely, only on their ratios).

For the delay-dependent priority system with preemption, we give a recursive form for W_p in terms of the W_i for the lower-priority messages.

Theorem 5.3[1]

For the delay-dependent priority system with preemption and $0 \le \rho < 1$,

$$W_p = \frac{1}{1 - \sum_{i=p+1}^{P} \rho_i \left(1 - \dfrac{b_p}{b_i} \right)} \left[\frac{W_0}{1-\rho} + \sum_{i=p+1}^{P} \frac{\rho_i}{\mu_p} \left(1 - \frac{b_p}{b_i} \right) \right.$$

$$\left. - \sum_{i=1}^{p-1} \frac{\rho_i}{\mu_i} \left(1 - \frac{b_i}{b_p} \right) - \sum_{i=1}^{p-1} \rho_i W_i \left(1 - \frac{b_i}{b_p} \right) \right] \tag{5.16}$$

This family is plotted in Fig. 5.5*a* and *b*.

[1] See Appendix C for proof of this theorem.

It is interesting to note the behavior of W_p as a function of ρ for the various disciplines discussed. The curves in Figs. 5.2 to 5.5 have been prepared to illustrate this behavior. The assumptions are that $\lambda_p = \lambda/P$ and $\mu_p = \mu$ ($p = 1, 2, \ldots, P$); for Figs. 5.4 and 5.5, $b_p = 2^{p-1}$. Of course, these special cases do not reveal the entire structure of the W_p, but they do give one an intuitive feeling about their general properties; the obvious reason for choosing these values is that they are easy to plot. Figures 5.2 and 5.3 show μW_p for the fixed-priority system without and with preemption. Figures 5.4 and 5.5 similarly show μW_p for the delay-dependent priority system. The curves shown are for $P = 2$ and $P = 5$. In addition, the case $P = 1$ is shown as a dashed curve in all the figures; clearly, for $P = 1$, $\mu W_p(\rho) = \rho/(1 - \rho)$ in all priority systems which corresponds to the strict first-come first-served discipline).[1] As such, the $P = 1$ case serves as a basis of comparison for all the curves.

Observe that, in general, the curves for the preemptive cases are more widely spaced than the corresponding curves for the nonpreemptive cases. Further, one can note that, in general, the curves for the fixed-priority system are more widely spaced than the corresponding curves for the delay-dependent priority system. Also note that, because of the rigid nature of the fixed-priority system, some of the curves for W_p extend beyond the value of $\rho = 1$. That is, although the service facility is saturated, only the lower-priority groups experience an infinite expected waiting time; some of the higher-priority groups have a finite expected wait under this overload condition. However, the delay-dependent priority system forces a fairly strong coupling (or interaction) among *all* the priority groups. Specifically, if any message remains in the queue for an extremely long time, it will eventually attain an extremely high value of priority; as such, it must eventually get served before any newly entering messages. Thus, if any group experiences an infinite expected waiting time, then they all do. This effect causes all the W_p curves to have poles at $\rho = 1$.

An important distinction between the two priority systems can be observed by considering the number of *degrees of freedom* in the specification of the systems. If we consider that the input traffic is specified, that is, P, λ_p, and μ_p ($p = 1, 2, \ldots, P$) are fixed (given) quantities, then we recognize that the fixed-priority system has *no* degrees of freedom left, and so the W_p are completely specified. This is not a desirable situation, since the system designer is not able to adjust the system's behavior. However, in the delay-dependent priority system, the variables b_p ($p = 1, 2, \ldots, P$) are at the disposal of the designer;

---

[1] The conservation law presented in Sec. 5.2 shows why $\mu W_p(\rho)$ for the case $P = 1$ must be independent of queue discipline.

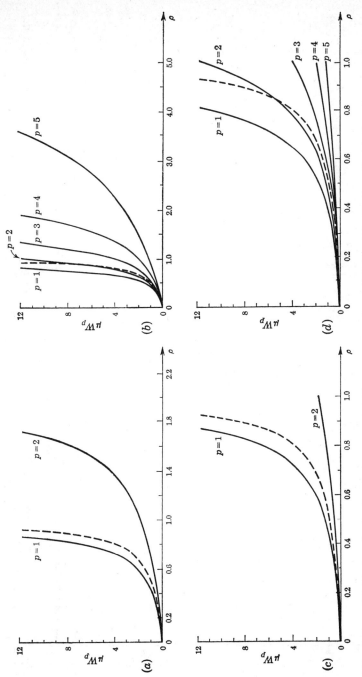

Fig. 5.2. $\mu W_p(\rho)$ for the fixed-priority system with no preemption. (a) $P = 2$; (b) $P = 5$; (c) $P = 2$, expanded scale; (d) $P = 5$, expanded scale.

79

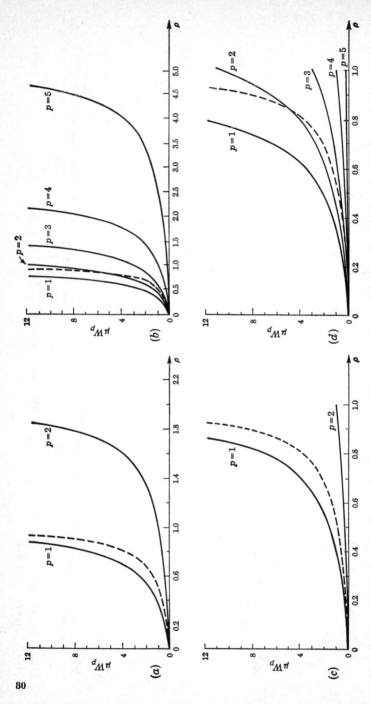

Fig. 5.3. $\mu W_p(\rho)$ for the fixed-priority system with preemption. (a) $P = 2$; (b) $P = 5$; (c) $P = 2$, expanded scale; (d) $P = 5$, expanded scale.

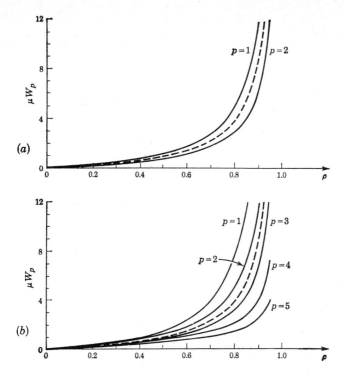

Fig. 5.4. $\mu W_p(\rho)$ *for the delay-dependent priority system with no preemption.* (*a*) $P = 2$; (*b*) $P = 5$.

this variability allows adjustment of the relative spacing of the W_p to a large degree. Finally, we note that if the higher-priority groups have shorter average message lengths than the lower-priority groups, then the average queue length is reduced.

5.2 A Conservation Law

An interesting phenomenon may be observed in the curves presented in Figs. 5.2 to 5.5. It appears that the curve for a strict first-come first-served system (dashed in all the figures) lies somewhere between the curves of the high- and low-priority messages. Perhaps a conservation law is at play here, holding constant some average value of the waiting times for the different priority groups. In fact, it is reasonable to expect such an invariance based on the simple physical argument that some messages are given preferential treatment and so

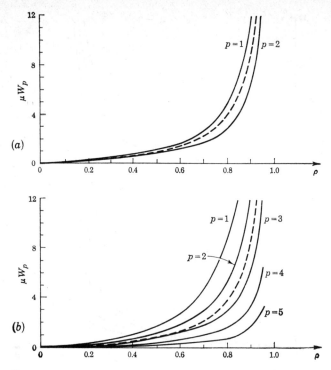

Fig. 5.5. $\mu W_p(\rho)$ *for the delay-dependent priority system with preemption.* (a) $P = 2$; (b) $P = 5$.

need not wait as long as they would in a first-come first-served system; as a result, low-priority messages are forced to wait some additional time.

Indeed, we find that there is a law of conservation which holds for the priority systems described; in fact, it holds for queueing systems subject to a large class of disciplines. A sufficient set of restrictions to define the class is as follows:

1. All messages remain in the system until completely serviced (i.e., no defections).
2. There is a single service facility which is always busy if there are any messages in the system.
3. Preemption is allowed only if the service-time distributions are exponential and the preemption is of the preemptive-resume type.
4. Arrival statistics are all Poisson, service statistics are arbitrary, and arrival and service statistics are all independent of each other.

For such a class, the conservation law states that, given a fixed set of arrival and service statistics, a particular weighted sum of the waiting times W_p is a *constant* independent of queue discipline. Once again, W_p is defined as the expected value of the time spent in the queue for a message with assigned parameter p.

Theorem 5.4[1]   (*the conservation law*)

For any queue discipline and any fixed arrival and service-time distributions subject to the above restrictions,

$$\sum_{p=1}^{P} \rho_p W_p = constant \text{ with respect to variation of queue} \quad (5.17)$$
$$\text{discipline}$$

where P represents the total number of groups to be distinguished in the traffic,[2] and where

$$\rho_p = \frac{\lambda_p}{\mu_p} = \text{[average arrival rate of } p\text{th group} = \lambda_p] \cdot \text{[expected duration of service time for a message from } p\text{th group} = 1/\mu_p]$$

In particular, for $\rho = \sum_{p=1}^{P} \rho_p$, we assert that

$$\sum_{p=1}^{P} \rho_p W_p = \begin{cases} \dfrac{\rho}{1-\rho} V_1 & 0 \le \rho < 1 \\ \infty & \rho \ge 1 \end{cases} \quad (5.18)$$

where $V_1 = \frac{1}{2} \sum_{p=1}^{P} \lambda_p E(t_p^2)$ \quad (5.19)

and $E(t_p^2)$ = second moment of service-time distribution for group p

V_1 may be interpreted as the expected time required to complete the message found in service upon entry, for a first-come first-served system. That is, convert the system at hand to one in which the same arrival and service-time distributions apply, but where the entire priority and preemptive structure is removed so that the system operates on a first-come first-served basis. Thus, V_1 is itself independent of the particular queue discipline chosen.

---

[1] See Appendix C for proof of this theorem. Along with the proof, two related corollaries are stated and proved.

[2] For an explicit description of the meaning of the p subscript, see the introductory remarks in Sec. 5.1. Roughly speaking, a higher p implies a higher priority.

Note that the conservation law constrains the allowed variation in the W_p for any discipline within the wide class considered. The conservation law states that the sum

$$\sum_{p=1}^{P} \frac{\lambda_p}{\lambda} W_p \tag{5.20}$$

(which weights the expected waiting time of the pth priority group by its relative arrival rate λ_p/λ) is a constant when all μ_p are equal. This sum (if multiplied by λ) represents the average number of messages in the queue (see Appendix C). The conservation law also states that the time-averaged waiting time[1]

$$\sum_{p=1}^{P} \frac{\lambda_p}{\lambda} \frac{1}{\mu_p} W_p \tag{5.21}$$

(which weights the W_p not only by λ_p/λ, but also by $1/\mu_p$, the average message length of a p type message) is a constant.

5.3   Time-shared Service

In this section we present results for a simple "round-robin" time-shared service facility and compare these results with a straightforward first-come first-served discipline. The round-robin discipline shares the desirable features of a first-come first-served principle and those of a discipline which services short messages first. Such a scheme is a likely candidate for the discipline of a large time-shared computational facility.

Let time be quantized into segments, each Q seconds in length. At the end of each time interval, a new message arrives in the system with probability λQ (result of a Bernoulli trial); thus, the average number of arrivals per second is λ. The service time of a newly arriving message is chosen independently from a geometric distribution such that, for $\sigma < 1$,

$$s_n = (1 - \sigma)\sigma^{n-1} \qquad n = 1, 2, 3, \ldots \tag{5.22}$$

where s_n is the probability that a message's service time is exactly n time units long (i.e., that its service time is nQ seconds).

The procedure for servicing is as follows: A newly arriving message

---

[1] Physically, we may think of this average as follows: Let us sample the system at random points in time; each time we sample, we record the time spent in the queue by the message currently being transmitted. The average value of this set of numbers is the average we are referring to.

joins the end of the queue and waits in line in a first-come first-served fashion until it finally arrives at the service facility. The server picks the next message in the queue and performs one unit of service upon it (i.e., services this message for exactly Q seconds). At the end of this time interval, the message leaves the system if its service (transmission) is finished; if not, it joins the end of the queue with its service partially completed, as shown in Fig. 5.6. Obviously, a message whose length is n time units long will be forced to join the queue n times in all before its service is completed.

Another assumption must now be made regarding the order in which events take place at the end of a time interval. We consider two types of systems. The first system allows the message in service to be ejected from the service facility (and then allows it to join the end of the queue if more service is required), and instantaneously after that a new message arrives (with probability λQ). We call this a *late-arrival* system (LAS). The second system reverses the order in which these events occur, giving rise to the *early-arrival* system (EAS). In both systems, a new message is taken into service at the beginning of a time interval.

First we consider the late-arrival system, which is similar to one considered by Jackson [50] for a different class of priority systems. By straightforward techniques, he arrives at the solution for the steady-state probability r_k that there are k messages in the system just before the time at which an arrival is allowed to occur (i.e., just after a message is ejected from service if there was a message in service); Jackson's result is

$$r_k = (1 - a)a^k \tag{5.23}$$

where
$$a = \frac{\rho\sigma}{1 - \lambda Q}$$

and
$$\rho = \frac{\lambda Q}{1 - \sigma}$$

This defines ρ, as usual, as the product of the average arrival rate λ and the mean service time $Q/(1 - \sigma)$. The notation of Jackson's result has

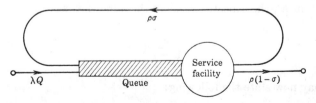

Fig. 5.6. *Round-robin time-shared service system.*

been altered to correspond to that used in this monograph. We may quickly obtain[1] the expected value E of the number k as

$$E = \frac{\rho\sigma}{1-\rho} \tag{5.24}$$

These results also apply to the time-shared service facility. For the time-shared system, we now state:

Theorem 5.5[1]

The expected value of the total time T_n spent in the late-arrival system by a message whose service is nQ seconds is

$$T_n = \frac{nQ}{1-\rho} - \frac{\lambda Q^2}{1-\rho}\left[1 + \frac{(1-\sigma\alpha)(1-\alpha^{n-1})}{(1-\sigma)^2(1-\rho)}\right] \tag{5.25}$$

where

$$\alpha = \sigma + \lambda Q$$

In Appendix C we demonstrate that $\alpha < 1$. An upper bound for T_n is easily obtained (by lower-bounding the quantity in brackets by unity) as

$$T_n \leq \frac{Q}{1-\rho}(n - \lambda Q) \tag{5.26}$$

We now consider the early-arrival system. Let r_k be the steady-state probability that there are k messages in the system just after an arrival is allowed to occur (i.e., just before the time at which a message is ejected from service if there is a message in service). In Appendix C it is shown that

$$r_k = \begin{cases} 1 - \rho & k = 0 \\ \dfrac{1-\rho}{\sigma}a^k & k = 1, 2, \ldots \end{cases} \tag{5.27}$$

where a and ρ are defined just as in the late-arrival system. From this, we obtain E', the expected value of the number k, as

$$E' = \frac{\rho}{1-\rho}(1 - \lambda Q) \tag{5.28}$$

We may now state the following:

---

[1] See Appendix C for proof of Eq. (5.24) and of Theorem 5.5.

Theorem 5.6[1]

The expected value T_n of the total time spent in the early-arrival system by a message whose service time is nQ seconds is

$$T_n = \frac{nQ}{1 - \rho} - \rho Q - \frac{\lambda Q^2 \rho}{1 - \rho} \left[1 + \frac{(1 - \sigma \alpha)(1 - \alpha^{n-1})}{(1 - \sigma)^2 (1 - \rho)} \right] \quad (5.29)$$

where α is defined as before.

An upper bound for T_n is easily obtained (by lower-bounding the quantity in brackets by unity) as

$$T_n \le \frac{Q}{1 - \rho} (n - \lambda Q \rho) - \rho Q \quad (5.30)$$

We now consider the case in which all messages wait for service in order of arrival and where, once in service, each message remains until it is completely serviced. It is easy to show that with T_n defined as before, we get:

Theorem 5.7[1]

The expected value T_n of the total time spent in the strict first-come first-served system by a message whose service time is nQ seconds is

$$T_n = \frac{QE}{1 - \sigma} + nQ \quad (5.31)$$

where

$$E = \frac{\rho \sigma}{1 - \rho}$$

Note that the distinction between the early- and late-arrival systems has disappeared, as, of course, it must. Note also that the expression defining E is the same as that in Eq. (5.24)—the average number of messages in the late-arrival system.

Let us now compare some of these results for time-shared systems. First we compare the values of E and E'. Let Δ be the difference between the expected numbers of messages in the early- and late-arrival systems. Then

$$\Delta = \frac{\rho}{1 - \rho} (1 - \lambda Q) - \frac{\rho}{1 - \rho} \sigma$$

and so

$$\Delta = \rho(1 - \sigma) = \lambda Q \quad (5.32)$$

---

[1] See Appendix C for proof of this theorem.

This result is quite reasonable, since for σ equal to zero (which is to say that each service time equals one time interval exactly) the difference Δ should be the probability of finding a message in the early-arrival system (which is merely ρ); for σ approaching unity, the difference approaches zero, since with probability $1 - \sigma$ a message will leave the system before (after) the next arrival. Note that Δ is always less than unity.

Now, if one wishes an *approximate* solution to the round-robin system, one might argue as follows: A message whose service time is nQ seconds must enter the end of the queue exactly n times. Roughly speaking (this approximation is evaluated presently), each time the message enters the queue, it finds E (or E') messages ahead of it. The time spent waiting for service each time around is then approximately QE. The time actually spent in service is exactly nQ. Thus, the approximation to T_n, which we label as T'_n, is

$$T'_n = nQE + nQ \tag{5.33}$$

Upon comparing this with Eq. (5.31) for the strict first-come first-served system in which

$$T_n = \frac{1}{1 - \sigma} QE + nQ$$

we see that for messages with length n less (greater) than $1/(1 - \sigma)$ the round-robin waiting time (for the late-arrival system) is less (greater) than that in the strict first-come first-served system. However, one notes that the average length (in service intervals) is exactly $1/(1 - \sigma)$. Thus, for this approximate solution, the crossover point for waiting time is at the mean number of service intervals. An evaluation of this approximation may be obtained by comparing the quantities T_n/Q as given in Eq. (5.25) and T'_n/Q as given in Eq. (5.33). That is, the approximation is only as good as the agreement between these two (for the late-arrival system[1]):

$$\frac{n}{1 - \rho} - \frac{\lambda Q x}{1 - \rho} \leftrightarrow \frac{n}{1 - \rho} - \frac{\lambda Q n}{1 - \rho} \tag{5.34}$$

where
$$x = 1 + \frac{(1 - \sigma\alpha)(1 - \alpha^{n-1})}{(1 - \sigma)^2(1 - \rho)}$$

[1] For the early-arrival system, we compare

$$\frac{n}{1 - \rho} - \frac{\lambda Q \rho x}{1 - \rho} - \rho \leftrightarrow \frac{n}{1 - \rho} - \frac{\lambda Q n}{1 - \rho}$$

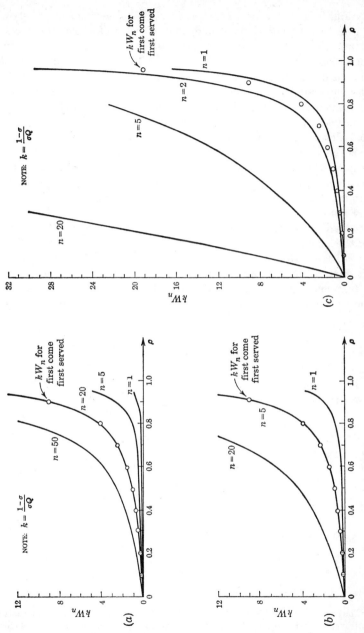

Fig. 5.7. $[(1 - \sigma)/\sigma Q]W_n(\rho)$ *for the late-arrival time-shared service system.* (a) $\sigma = 19/20$; (b) $\sigma = 4/5$; (c) $\sigma = 1/5$.

NOTE: $k = \dfrac{1-\sigma}{\sigma Q}$

89

In Fig. 5.7a to c, curves of $[(1 - \sigma)/\sigma Q]W_n(\rho)$ are plotted[1] to show the general behavior of the round-robin structure for the late-arrival system. Points corresponding to the first-come first-served case have been included on each graph. The normalization factor $(1 - \sigma)/\sigma Q$ was used so that in these figures, as well as in Figs. 5.2 to 5.5, the same first-come first-served curves, $\rho/(1 - \rho)$, would appear. Further, it is important to note that μ, the average service rate, equals $(1 - \sigma)/Q$ in the discrete case, and thus Fig. 5.7 plots $(\mu/\sigma)W(\rho)$, whereas Figs. 5.2 to 5.5 plot $\mu W(\rho)$. Note that the only parameter change in Fig. 5.7a to c is the value of σ.

Figure 5.7a to c indicates the accuracy of the approximation, discussed above, in which the crossover point for waiting times is at the mean number of service intervals, $1/(1 - \sigma)$. In Fig. 5.7a and b there is no noticeable difference (on the scale used) between the first-come first-served points and the curve for $n = 1/(1 - \sigma)$; moreover, in Fig. 5.7c the points fall between the curves for $n = 1$ and $n = 2$, since $1/(1 - \sigma) = 1.25$.

It is interesting to note that both round-robin disciplines, along with the first-come first-served (FCFS) discipline, offer an example of the validity of the conservation law [Eq. (5.17)]. That is, if we define

$$T_n(\text{FCFS}) \text{ as given by Eq. (5.31)}$$
$$T_n(\text{LAS}) \text{ as given by Eq. (5.25)}$$
$$T_n(\text{EAS}) \text{ as given by Eq. (5.29)}$$

and $$W_n(\cdot) = T_n(\cdot) - nQ$$

then it is a simple algebraic exercise to show that

$$\sum_{n=1}^{\infty} \rho_n W_n(\cdot) = \text{constant} = \frac{Q\rho^2\sigma}{(1 - \rho)(1 - \sigma)} \tag{5.35}$$

where $$(\cdot) = (\text{FCFS, LAS, EAS})$$
and $$\rho_n = \rho s_n = \rho(1 - \sigma)\sigma^{n-1}$$

But $\lambda_n = \lambda s_n = $ average arrival rate of units which require n service intervals

Thus, Eq. (5.35) may be rewritten as

$$W \equiv \sum_{n=1}^{\infty} \frac{\lambda_n}{\lambda} W_n(\cdot) = \text{constant} = \frac{Q\rho\sigma}{(1 - \rho)(1 - \sigma)}$$

[1] The expected value W_n of the time spent in the queue for a unit which requires n intervals of service may be obtained from T_n by means of the relationship $W_n = T_n - nQ$.

This equation states that the mean waiting time[1] W is a constant which is independent of the three disciplines discussed. Thus, as is to be expected, one does not improve the mean wait W; however, by introducing the round-robin system analyzed in this study, one manipulates the relative waiting time for different messages (while maintaining a constant W) and thus imposes a time-sharing system which gives preferential treatment to short messages.

5.4 Conclusions and Extensions

Let us consider some extensions of the material presented in this chapter. Having discussed the fixed-priority system and the delay-dependent priority system separately, it seems natural to consider a combination of the two. In particular, consider a discipline in which a message from priority group p entering the queue at time T is assigned two numbers: a_p and b_p. The priority $q_p(t)$ at time t associated with that message is calculated as

$$q_p(t) = a_p + (t - T)b_p \qquad (5.36)$$

where the range of allowed t is from T up to the time at which the message's service is completed. This priority scheme takes on many interesting and familiar forms in certain special cases.[2] Of course, when b_p is identically zero, we have the fixed-priority system, and when a_p is identically zero, we have the delay-dependent priority system.

Because of the importance of the form of Eq. (5.36), a solution for the general case could be very valuable. This has not been accomplished, but an attack on the case of fixed (but nonzero) b_p and variable a_p has been made by Jackson [50]. He considered a model in which time was quantized; during each time interval, a new message arrived with probability λQ; if the system was nonempty, a completion of service occurred with probability $1 - \sigma$. Among his results are the following bounds on the equilibrium mean waiting time W_p for messages with priority number a_p (the notation of his result has been altered to correspond to that used in this monograph, i.e., the larger

---

[1] This mean waiting time is an appropriate average of W_n, since it weights W_n by the fractional number of messages λ_n/λ which must suffer that waiting time.

[2] For example, a_p and b_p may be chosen so as to describe the following priority disciplines: first-come first-served, last-come first-served, random ordering of service, and mixtures of the above.

the a_p the higher the priority):

$$W \frac{1 - \rho}{1 - \sum\limits_{i=p}^{P} \rho_i} \leq W_p \leq \frac{W}{1 - \cdot \sum\limits_{i=p+1}^{P} \rho_i} \tag{5.37}$$

where
$$W = \frac{\sigma}{1 - \sigma} \frac{\rho}{1 - \rho}$$

$$\rho = \frac{\lambda Q}{1 - \sigma}$$

$$\rho_i = \rho P_r \text{ [entering message is assigned}$$
$$a_i \text{ as parameter, where } a_i = i]$$

Jackson also goes into some detail for the case of $P = 2$ (i.e., only two different priority classes) and derives certain expressions, in matrix notation, for the average waiting times for the two priorities; values for these expressions are tabulated in his appendix [50]. In [51] he derives the asymptotic behavior of the waiting-time distribution for this class of dynamic priority queueing models.

Recently, at an informal lecture at M.I.T., John D. C. Little analyzed a very interesting problem associated with priority queueing. He considered the case in which arrivals to the system from the pth group were Poisson at an average rate of λ_p. Associated with such arrivals were a mean service time $1/\mu_p$ and a cost to the server of C_p dollars for every second that each arrival from the pth group remained in the system (queueing time plus service time). He then solved for that priority discipline which minimized the total time-averaged cost to the server. He found that the solution was a fixed-priority system in which the highest priority was given to that group with the largest product $\mu_p C_p$. Specifically, he reordered the subscripts such that

$$\mu_1 C_1 \leq \mu_2 C_2 \leq \cdots \leq \mu_P C_P \tag{5.38}$$

where, again, the larger the subscript, the higher the priority. This is a most interesting result, worthy of careful attention.

In reviewing the theorems of this chapter, we find it appropriate to state the important conclusions once again. The first conclusion we would like to emphasize is the versatility inherent in a delay-dependent priority structure.[1] By this we mean that a system designer has at his

---

[1] Recently, under the author's direction, R. P. Finkelstein studied a more general case of delay-dependent priority systems, the results of which are reported upon in his Master's thesis entitled "Study of Time Dependent Priority Schemes," presented to the Department of Electrical Engineering at the Massachusetts

disposal a whole set of parameters (the set b_p) with which he can adjust the relative waiting times W_p. He must have this freedom if he intends to satisfy, or come close to satisfying, a set of specifications given him by the intended user of the system. In general, the user will specify the traffic to be handled; i.e., he will specify the number P of priority groups and the average arrival rate λ_p and average message length $1/\mu_p$ for each of these groups.[1] Then the user will specify a set of *relative* W_p that he desires from the system. The additional number of degrees of freedom available to the designer from the set b_p is just what is necessary to satisfy the user's demands. Without this freedom the set W_p is fully determined (as in the fixed-priority system) and the designer cannot alter their relative values. Even with the b_p, certain limitations exist. First, the function W_P cannot lie below W_P for the fixed-priority system, since in the fixed-priority system the Pth group is given complete priority over all other groups, and members of this group interfere only with each other. Secondly, the conservation law clearly puts a constraint on the absolute values of the set W_p.

The conservation law, although proven for the class of systems described in this chapter, probably holds for a more inclusive class. Indeed, we have seen one case of a queue discipline (the time-shared service system) which falls outside our defined class and which nevertheless obeys the conservation law.

At least two interesting conclusions can be drawn from the conservation law. Firstly if all $\mu_p = \mu$, then the law states that a meaningful average[2] of the average waiting times is invariant with respect to a change in queue discipline. Therefore, one need not search for a special queue discipline in the class to minimize this average—it is fixed. Secondly, if the μ_p are arbitrary, then the time-averaged[3] waiting time is invariant to a change in the queue discipline.

Finally, we would like to refer the reader to Chap. 7, in which are described the results of certain simulation experiments. These experiments demonstrate that the conservation law holds for a fixed-priority

---

Institute of Technology on Aug. 19, 1963. In this new system a message from the pth priority group entering the queue at time T has associated with it a priority $q_p(t)$ at time t calculated from $q_p(t) = b_p'(t - T)^n$. The results of that study show that any such nth order delay-dependent priority system may be reduced to our first-order system ($n = 1$) by choosing a new $b_p = (b_p')^{1/n}$.

[1] Note that when λ_p and $1/\mu_p$ are specified, $\rho = \sum_{p=1}^{P} \lambda_p/\mu_p$ is also specified.

[2] This average weights the waiting time by the relative number of messages which must suffer that waiting time.

[3] See footnote on page 84.

system in a communication network in which $\mu_p = \mu$.   The results are
tabulated in Table 5.1.

Table 5.1.   *Values of μCT for the star net obtained from simulation, demonstrating* experimental evidence of the conservation law.*

$\dfrac{\gamma}{\mu C}$	Traffic matrix	μCT					
		Identical capacity		Proportional capacity		Square root capacity	
		$P=1$	$P=3$	$P=1$	$P=3$	$P=1$	$P=3$
.0625	τ_1	49.6	49.2	48.4	47.6	37.3	37.2
.0625	τ_3	49.6	50.4	51.2	50.8	50.0	50.8
.250	τ_1	503.	490.	73.5	73.8	61.2	61.2
.250	τ_3	84.5	82.7	87.3	85.2	86.0	82.8
.376	τ_1			103.	111.	100.	92.
.376	τ_3	154.	159.	151.	153.	144.	154.

* The value of P, as before, gives the number of priority groups in the fixed-priority system which was simulated. C refers to the total capacity assigned to the net.   For the definition of other terms in the table, the reader is referred to Chap. 7.

chapter 6

Random Routing Procedures

Random routing procedures are those decision rules in which the choice as to the next node to visit is made according to some probability distribution over the set of neighboring nodes.[1] In this chapter we connect a group of nodes to each other and apply a random routing procedure to the resultant net. Two results of interest emerge from this investigation: First, we find that for a particular class of random routing procedures, we are able to solve for the expected number of steps that a message must take before arriving at its destination. Second, we derive an expression for the expected time that a message spends in the net. In the analysis, we use our model with the inclusion of the independence assumption (see Chap. 3). A number of results tangent to the main discussion may be found in Appendix D.

One may reasonably ask why random routing procedures are of interest. Their main advantage is that they are simple, both in conception and in realization in a practical system. Another advantage is that systems operating under a random routing procedure are relatively insensitive to changes in the structure of the network; i.e., if some of the channels disappear, the routing procedure continues to function without considerable degradation in performance. Moreover, since the random routing procedure does not make use of directory information, changes in the network structure need not be made known to all

---

[1] For example, one random routing procedure may be defined such that the next node to visit is chosen with equal probability from the set of idle channels leading out of the present node.

the nodes.   This fact becomes increasingly important in a hostile or fluid environment in which changes in the network take place continuously.   If it were necessary to transmit information around the network informing all nodes of each change, the network might easily become flooded with directory information alone, thus leaving no transmission capability for message traffic.   Furthermore, it may be added that random routing procedures offer examples of some of the few routing procedures in which it is possible to get some meaningful analytic results.   Thus, they may well serve as a measure of the quality of performance of other routing procedures.

Clearly, there are a number of disadvantages inherent to random routing procedures.   The major difficulty is that the procedure does not take advantage of certain available information.   In particular, the topology of the network along with the destination of a message suggest that certain paths are to be preferred over others; the random routing procedure neither recognizes nor utilizes this information.   As a result, messages are forced to follow a random path.   When finally a message is fortunate enough to be transmitted to its destination, it is dropped from the network.   In 1962, R. Prosser [40] offered an approximate analysis of a random routing procedure in a communication net in which he showed that such procedures are highly inefficient in terms of message delay but extremely stable (i.e., they are relatively unaffected by small changes in the network structure).

The overall effect of random routing is to increase the internal traffic that the network is required to handle; consequently, the external traffic that may be applied to the net is greatly reduced.   In addition, the time that a message spends in the network is increased, thus reducing the grade of service to the user of the system.

6.1   Markov Model: Circulant Transition Matrices

The two quantities of most interest to the user in the study of any routing procedure are the expected time T that a message spends in the net and the mean total traffic that the network can handle.   In the following discussion, we center our interest on the first of these; as we shall see, the analysis yields an answer for the mean total traffic as well.

The time that a message spends in the system is the sum of the time spent at each node that the message visits.   If it turns out that the expected time $T(n)$ spent at node n is approximately the same value T_0 for all n, then it is clear that

$$T \approx \bar{n}T_0 \tag{6.1}$$

where $\bar{n}$ is the average number of steps[1] that a message must take to reach its destination.

In the case of random routing procedures, the calculation of $\bar{n}$ may be carried out independently of the calculation of T_0, whereas the converse is not necessarily true. We therefore concentrate initially upon the calculation of $\bar{n}$. Note that this approach effectively removes from consideration all questions of queueing and leaves only the matters of topology and routing. We return to the queueing aspect of the problem in Sec. 6.4.

The model with which we choose to represent the random routing procedure is that of a Markov process with $N + 1$ states. Each node in the network corresponds to a distinct state, where the nodes are numbered $0, 1, 2, \ldots, N$. Associated with each message is an originating node and a destination. The node of origination will be the *initial position* of the message in the process, and the destination will be the *absorbing node* for the message. The routing procedure is reflected in the Markov model as the one-step transition probability matrix P. A typical entry p_{ij} in this matrix is then the probability that a message in node i will be transmitted to node j next,[2] where, as usual, we require that $p_{ij} \geq 0$. The matrix P, together with the a priori distribution of the originating node, completely describe the Markov process.[3]

The analytic solution for $\bar{n}$ for an arbitrary matrix P cannot in general be obtained in closed form. A number of different expressions for $\bar{n}$ which involve infinite summations are well known and are summarized in Appendix D. These open forms are of limited utility for analytical work. In order to obtain a closed-form solution, one must put some structure into the matrix P. The added structure should not be such that the resultant solution, although elegant and concise, is of no use in the problem at hand.

Thus, we find ourselves in the position of defining a restricted class of Markov processes (that is, P matrices) which are analytically tractable and at the same time useful in answering questions about interesting random routing procedures.

As it turns out, there is a subclass of Markov processes which includes a large number of very interesting random routing procedures and

---

[1] The average number of nodes visited is $\bar{n} + 1$, but the last node visited is the destination itself, and by convention we agree that once the destination is reached, the message is immediately dropped from the net. The quantity of interest is therefore the average number of steps taken (also referred to as the *average path length*).

[2] Of course, this probability is conditioned on the fact that the destination for this message is not node i itself.

[3] Furthermore, the network topology, or connectivity, is implicitly described by the P matrix; i.e., channels exist only between those nodes i, j for which $p_{ij} > 0$.

which yields to analysis.   The set of Markov processes included in this class is defined to be those whose P matrices are of the following form:

$$P = \begin{bmatrix} q_0 & q_1 & q_2 & \cdots & q_N \\ q_N & q_0 & q_1 & \cdots & q_{N-1} \\ q_{N-1} & q_N & q_0 & \cdots & q_{N-2} \\ \cdot & \cdot & \cdot & \cdots & \cdot \\ q_1 & q_2 & q_3 & \cdots & q_0 \end{bmatrix} \qquad (6.2)$$

This particular form of transition matrix is known as a *circulant* matrix, and it describes a class of Markov processes referred to as *cyclical random walks*.   As in any probability transition matrix, each row must sum to unity; in this case, that constraint may be expressed as

$$\sum_{i=0}^{N} q_i = 1$$

Furthermore, we require that the Markov chain described by P in Eq. (6.2) is irreducible[1] (i.e., each state can be reached from all other states).

Note that each row of this matrix is the same as the row above it except for a rotation of one place to the right.   One major feature of this matrix is that no matter which node (row) one chooses, the remainder of the net (matrix) appears identical; i.e., looking out into the rest of the net, each node sees the same topological structure.

In summary, then, we consider those random routing procedures which are describable by finite irreducible Markov processes whose probability transition matrices are circulant matrices, i.e., routing procedures which are cyclical random walks over the space of nodes (communication centers).

6.2   The Average Path Length

As shown in Appendix D, the cyclical random walk yields the following expression for the expected number of steps $\bar{n}_i$ required to travel from node i to node N (note that the choice of node N as the destination for each message is in no way restrictive, since the choice of i is arbitrary, and the entries in the matrix can easily be relabeled):

Theorem 6.1

The average path length $\bar{n}_i$ from node i to node N for any finite-dimensional irreducible Markov process whose probability transi-

_____

[1] See Feller [18, Sec. XV.4].

tion matrix is a circulant matrix [see Eq. (6.2)] is

$$\bar{n}_i = \sum_{r=1}^{N} \frac{1 - \theta^{r(i+1)}}{1 - \sum\limits_{s=0}^{N} q_s \theta^{sr}} \tag{6.3}$$

where $i = 0, 1, 2, \ldots, N$, where θ is the $(N + 1)$th primitive root of unity, i.e.,

$$\theta = e^{2\pi j/(N+1)} \tag{6.4}$$

and where

$$j = \sqrt{-1}$$

This result is rather simple and is easily evaluated for any choice of the set q_s ($s = 0, 1, 2, \ldots, N$). In addition, the *generating function*[1] $F_{iN}(t)$ for the first passage time from node i to node N takes the form

$$F_{iN}(t) = \frac{1 + (1 - t) \sum\limits_{r=1}^{N} \dfrac{\theta^{r(i-N)}}{1 - t \sum\limits_{s=0}^{N} q_s \theta^{sr}}}{1 + (1 - t) \sum\limits_{r=1}^{N} \dfrac{1}{1 - t \sum\limits_{s=0}^{N} q_s \theta^{sr}}} \tag{6.5}$$

One may ask about the behavior of cyclical random walks when certain averages are taken over the index i (which represents the originating node in the model). The simplest assumption to make is that the a priori probability of starting a message in any node is uniform over the set of possible nodes ($i = 0, 1, 2, \ldots, N - 1$), where we clearly do not allow a message to originate at its destination. In this case, we define $\bar{n}$ to be the average over i of the number of steps required to reach the destination, and we obtain the following result (as shown in Appendix D):

$$\bar{n} = \frac{1}{N} \sum_{i=0}^{N} \bar{n}_i = \frac{N+1}{N} \sum_{r=1}^{N} \frac{1}{1 - \sum\limits_{s=0}^{N} q_s \theta^{sr}} \tag{6.6}$$

We digress, for a moment, to consider an interesting alternative assumption which may be made about the a priori distribution of originating nodes. In particular, let us assume that the probability of starting a message at node i is equal to q_{i+1} for $i = 0, 1, \ldots,$ $N - 1$; for consistency, we allow a message to originate at node N with

---

[1] See Appendix D for the definition of generating function.

probability q_0. Defining $\bar{n}'$ as the result of this averaging process, we find[1]

$$\bar{n}' = \sum_{i=0}^{N-1} q_{i+1}\bar{n}_i + q_0\bar{n}_N = N \qquad (6.7)$$

This result is surprisingly simple. It states that if one chooses to originate messages over the set of nodes in the manner just described, then, no matter what the actual choice of the set q_i (subject to the finiteness and irreducibility of P), the average number of steps necessary to reach node N (the destination) turns out to be N.*

6.3 Examples of Networks Included by the Model

Some of the most commonly considered network configurations are included by our model of a cyclical random walk. For example, the well-known *ring* configuration corresponds to the matrix P and topological diagram shown in Fig. 6.1. Further, the *nearest-neighbor* configuration is included and corresponds to the matrix P and the diagram shown in Fig. 6.2.

Let us now consider a more general type of nearest-neighbor configuration which is included in our model, namely, one in which an

---

[1] See Appendix D for proof.

* Professor C. E. Shannon, (M.I.T.) in a private discussion, has given a very nice explanation for the simplicity of this result. Essentially, his comment was that since each node in the network is equivalent, the average number of steps required to return to a node must be $N + 1$ (the number of nodes in the system). Now, if we assume that we begin at node N (the destination) and make one step out into the network, we find ourselves at node i with probability q_{i+1} and at node N with probability q_0, which is just the a priori distribution described above. Since we have already taken one step, the average number of steps left to return to node N must now be $(N + 1) - 1$, which is merely N [as stated by Eq. (6.7)].

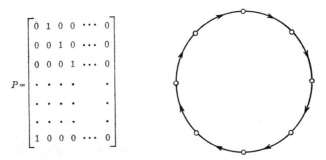

$$P = \begin{bmatrix} 0 & 1 & 0 & 0 & \cdots & 0 \\ 0 & 0 & 1 & 0 & \cdots & 0 \\ 0 & 0 & 0 & 1 & \cdots & 0 \\ \cdot & \cdot & \cdot & \cdot & & \cdot \\ \cdot & \cdot & \cdot & \cdot & & \cdot \\ \cdot & \cdot & \cdot & \cdot & & \cdot \\ 1 & 0 & 0 & 0 & \cdots & 0 \end{bmatrix}$$

Fig. 6.1. *Ring-net configuration.*

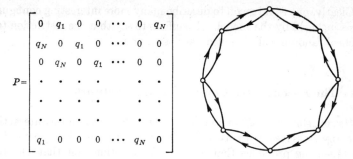

$$P = \begin{bmatrix} 0 & q_1 & 0 & 0 & \cdots & 0 & q_N \\ q_N & 0 & q_1 & 0 & \cdots & 0 & 0 \\ 0 & q_N & 0 & q_1 & \cdots & 0 & 0 \\ \cdot & \cdot & \cdot & \cdot & & \cdot & \cdot \\ \cdot & \cdot & \cdot & \cdot & & \cdot & \cdot \\ \cdot & \cdot & \cdot & \cdot & & \cdot & \cdot \\ q_1 & 0 & 0 & 0 & \cdots & q_N & 0 \end{bmatrix}$$

Fig. 6.2. *Nearest-neighbor configuration.*

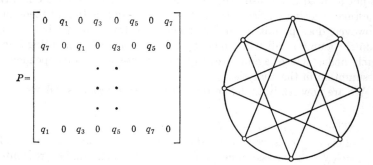

$$P = \begin{bmatrix} 0 & q_1 & 0 & q_3 & 0 & q_5 & 0 & q_7 \\ q_7 & 0 & q_1 & 0 & q_3 & 0 & q_5 & 0 \\ & & & \cdot & \cdot & & & \\ & & & \cdot & \cdot & & & \\ & & & \cdot & \cdot & & & \\ q_1 & 0 & q_3 & 0 & q_5 & 0 & q_7 & 0 \end{bmatrix}$$

Fig. 6.3. *A more general nearest-neighbor configuration.*

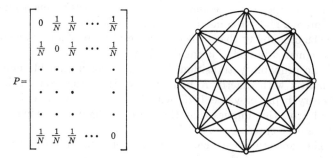

$$P = \begin{bmatrix} 0 & \frac{1}{N} & \frac{1}{N} & \cdots & \frac{1}{N} \\ \frac{1}{N} & 0 & \frac{1}{N} & \cdots & \frac{1}{N} \\ \cdot & \cdot & \cdot & & \cdot \\ \cdot & \cdot & \cdot & & \cdot \\ \cdot & \cdot & \cdot & & \cdot \\ \frac{1}{N} & \frac{1}{N} & \frac{1}{N} & \cdots & 0 \end{bmatrix}$$

Fig. 6.4. *Fully connected net.*

arbitrary set of links emanates from each node (the same set for each node).   An example of this type of network is shown in Fig. 6.3.[1]

The *fully connected* configuration with uniform probabilities, a network of considerable interest, is also included by the model (Fig. 6.4).

---

[1] An undirected link indicates a two-way connection.

Clearly, one could go on to describe many more interesting configurations included by the model. It suffices to say that the restriction to a cyclical random walk is not a very severe one.

6.4 The K-connected Communication Network

Having solved for $\bar{n}$, we now consider the dynamics of the interaction of many messages in a network with random routing (i.e., we consider the queueing problems that arise). We mention first that Theorem D.1 states that, in general, an arbitrary transition probability matrix is not allowed if we insist that no channels be idle when there is any message awaiting service in the queue. The corollary to that theorem, however, does permit a particular choice of transition probabilities, namely, a uniform distribution over the set of channels emanating from each node. Once again, we require the use of the independence assumption in the analysis.

We are now ready to define the *K-connected network* as follows:

1. The network contains $N + 1$ nodes.
2. Poisson message-arrival statistics with an average rate of $\gamma/(N + 1)$ messages per second arrive from external sources into each node.
3. Message lengths are exponentially distributed with mean length $1/\mu$ bits per message and are newly chosen each time a message enters a node (the independence assumption).
4. The probability that node k is the destination of a message which originated in node j is equal to $1/N$ for all $k \neq j$.
5. The probability transition matrix P is an irreducible circulant matrix with exactly K nonzero entries in each row, each such entry being of value $1/K$, with all diagonal terms equal to 0. Clearly, this implies that there are K channels leaving and K channels entering each node.
6. The capacity of each channel in the net is equal to $C/K(N + 1)$. Thus, the total channel capacity of the net is C.

We note that items 2 and 4 imply that the traffic matrix is a uniform matrix with each off-diagonal entry equal to $\gamma/N(N + 1)$. Clearly, with these definitions, all nodes behave statistically in the same way. Accordingly, let P_n be the probability that there are n messages in a particular node (any node is representative of all nodes) in the steady state. We then state:

Theorem 6.2[1]

For the K-connected net,

$$
P_n = \begin{cases} P_0(\bar{n}\rho)^n \dfrac{K^n}{n!} & n = 0, 1, 2, \ldots, K \\[2ex] P_0(\bar{n}\rho)^n \dfrac{K^K}{K!} & n \geq K \end{cases} \tag{6.8}
$$

provided $\bar{n}\rho < 1$, where

$$
\rho = \frac{\gamma}{\mu C} \tag{6.9}
$$

$$
P_0 = \left[\sum_{n=0}^{K-1} \frac{(K\bar{n}\rho)^n}{n!} + \frac{(K\bar{n}\rho)^K}{(1 - \bar{n}\rho)K!} \right]^{-1} \tag{6.10}
$$

$$
\bar{n} = \frac{N + 1}{N} \sum_{r=1}^{N} \frac{1}{1 - (1/K) \sum_{s \in S} \theta^{sr}} \tag{6.11}
$$

$\theta = e^{2\pi j/(N+1)} = (N + 1)$th primitive root of unity $\qquad$ (6.12)
S = set of integers which correspond to positions of nonzero elements of first row of P

Note that P_n in the K-connected net behaves very much like P_n in the multiple-channel system [see Eq. (A.9)]; the only difference is the introduction of $\bar{n}$.

Further, we can solve for T, the expected time that a message spends in the network (as defined in Sec. 6.1):

Theorem 6.3[2]

For the K-connected net,

$$
T = \frac{(N + 1)K\bar{n}}{\mu C} + \frac{\bar{n}(N + 1)}{\mu C(1 - \bar{n}\rho)} \frac{1}{(1 - \bar{n}\rho)S_K + 1} \tag{6.13}
$$

where
$$
S_K = \sum_{n=0}^{K-1} (K\bar{n}\rho)^{n-K} \frac{K!}{n!} \tag{6.14}
$$

and ρ and $\bar{n}$ are defined in Theorem 6.2.

[1] This theorem (proved in Appendix D) is similar to one given by Jackson [37]. The differences are the introduction of the notion of a destination and the presence of an explicit expression for the geometric decay factor $\bar{n}\rho$.

[2] See Appendix D for proof of this theorem.

The first term in the expression for T is merely the expected time that a message spends in transmission between nodes; the second term is the expected time that a message spends waiting on queues as it passes through the net.

In Sec. 6.1, it was stated that the mean total traffic that the network could handle would be found in the course of our investigation of T. Indeed this is so, as can be seen by examining Theorem 6.2. The condition $\bar{n}\rho < 1$ is merely a statement of the fact that the mean total traffic that can be accepted at each node from external sources must be less than $C/\bar{n}(N + 1)$ bits/sec.

6.5   Conclusions

Certain aspects of a class of random routing procedures have been investigated in this chapter. One of the main results was a solution for $\bar{n}$, the expected number of steps that a message takes in a network incorporating a random routing procedure (describable by a finite-dimensional irreducible circulant transition matrix). This includes a large number of interesting random routing procedures. For such nets,

$$\bar{n} = \frac{N + 1}{N} \sum_{\tau=1}^{N} \frac{1}{1 - \sum_{s=0}^{N} q_s \theta^{sr}} \tag{6.6}$$

Furthermore, we found that if we restrict our class of nets to K-connected networks into which we introduce the independence assumption, then T, the expected time that a message spends in the net, is

$$T = \frac{(N + 1)K\bar{n}}{\mu C} + \frac{\bar{n}(N + 1)}{\mu C(1 - \bar{n}\rho)} \frac{1}{(1 - \bar{n}\rho)S_K + 1} \tag{6.13}$$

We are now in a position to evaluate the performance of random routing procedures. Table 6.1 gives the value of $\bar{n}$ for the K-connected net of 13 nodes at four[1] different values of K. Two sets of numbers are presented: the first set represents $\bar{n}$ for the case of random routing, and the second represents $\bar{n}$ for the set of shortest paths (a fixed routing procedure) in the same network. As stated in the introductory remarks to this chapter, the number of nodes visited by each message is considerably larger for random routing procedures than for simple

---

[1] We do not include $K = 1$, since this defines the ring net, which is not truly a net with random routing.

Table 6.1. *Average number of steps for random routing and for the set of shortest paths in the K-connected net.*

K	Net*	$\bar{n}$ for random routing	$\bar{n}$ for the set of shortest paths
2		30.3	3.5
3		14.65	2.25
4		14.0	1.67
12		12.0	1.0

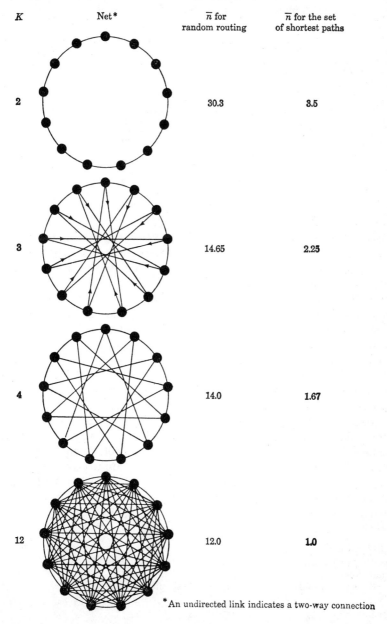

*An undirected link indicates a two-way connection

Table 6.2.   *Values of μCT for random routing (calculated) and for fixed routing (simulated) in a K-connected net for which $K = 4$.*

$\rho = \dfrac{\gamma}{\mu C}$	μCT random routing	μCT fixed routing
1/128	728	87.8
1/64	731	88.4
1/32	774	93.7
1/16	1803	102.
1/8	Network overloaded	120.
1/4	Network overloaded	173.
1/2	Network overloaded	580.

fixed routing procedures.[1]   Furthermore, since the maximum traffic that can be accepted at each node is on the order of $C/\bar{n}(N + 1)$ bits/sec, and since $\bar{n}$ is excessively large, random routing also reduces the total traffic that a net can sustain.

A digital simulation[2] was run for the K-connected net with $K = 4$ and $N + 1 = 13$ for a number of different traffic loads.   The routing procedure was fixed, and the quantity measured was T, the average time spent in the net.   Table 6.2 presents the results of this experiment along with the results for random routing calculated from Eq. (6.13). Both sets of numbers are for identical externally applied traffic statistics.   It is immediately obvious from this table that random routing procedures are extremely costly in terms of message delay.

In summary, we note that the price one pays for random routing is threefold: the number of nodes visited by each message is large, the time spent at each node is large, and the mean total traffic that the network can accept is small (all these comments are relative to a fixed routing procedure as described in Chapter 7).   The advantages of random routing are that the procedures are simple, little information is required about the rest of the network, and the system degrades slowly in the presence of a hostile or fluid environment.

---

[1] By a fixed routing procedure we mean one in which, given a message's origin and destination, there exists a unique path through the net which the message must follow.

[2] See Appendix E for a description of the simulation program.

chapter **7**

Simulation of
Communication Nets

In this chapter we describe the results obtained with a communication-net simulation program. The simulation runs were made on Lincoln Laboratory's TX-2 high-speed digital computer [52]. The purpose of the simulation was to test experimentally the results and predictions obtained from our theoretical investigation. Furthermore, as anticipated, the results of the experimentation suggested new lines of study. Thus, the simulation served a threefold function: first, it offered a means of verification for theoretically obtained results; second, it allowed experimentation in areas in which the mathematical theory was unmanageable;[1] and third, it acted as a feedback mechanism by suggesting new ideas.

An operational description of the simulation program may be found in Appendix E. In Sec. 7.1, we describe the nets that were simulated. The specific form of alternate routing used is also defined in that section. The results of our experimentation are presented graphically in Sec. 7.2. Of interest in the experimentation were the effects of network topology, channel capacity assignment, and alternate routing on the average message delay. In Sec. 7.3, we present two theorems related to the effects and design of alternate routing procedures.

---

[1] It also allowed us to test the validity of the independence assumption.

7.1 Specific Network Descriptions

In Appendix E are listed the types of data required to specify a communication net. In the present section, we give some specific data which define the nets that were simulated. To achieve this end, we must describe the format in which the data are presented. In particular, we define an *incidence matrix* I in which an entry of unity in the ij position indicates the presence of a channel leading from node i to node j $(i, j = 1, 2, \ldots, N)$ and the absence of an entry indicates the absence of that channel. Note that the matrix I therefore defines the topological structure of the net. The exact channel capacity assigned to each channel need not be specified, since, if desired, the program will automatically allocate capacity according to one of three assignments: square root, proportional, and identical capacities.[1] Of course, the total capacity assigned to the net must be specified separately.

The traffic matrix τ is defined such that an entry in the ij position gives the *relative* traffic which has node i as origin and node j as destination. The total traffic γ must be given separately; it defines the average number of messages entering the network per second.

Furthermore, the routing procedure must be specified. We choose first to define a specific form of *alternate routing procedure;* we then define a special case of alternate routing as *fixed routing.* Our alternate routing procedure is defined to be a decision rule which operates on a set of lists L_{ij} as follows: Given that a message is now in node i and has, for a final destination, node j, the decision rule consults list L_{ij} in order to decide which node the message will visit next. The message is routed to the first node entered on this list if the transmission channel connecting node i to this node is currently idle; if this channel is busy, the second entry on the list is tried, etc., until the list is exhausted, in which case, the message must wait on a queue in node i until a transmission channel connected to any node on the list becomes available at node i.[2] Clearly, this form of routing procedure offers many choices at each step; the lists express a preference of one path over another. Indeed, this is a simple system which offers alternative routes under busy conditions. It is convenient to write this set of lists in the form of a matrix R, the ijth entry being a list of numbers such that the first

---

[1] Recall that the capacity assignment described by Eq. (2.4) is referred to as *square-root* assignment; assignment according to Eq. (2.8) is referred to as *proportional* assignment. We now define a capacity assignment wherein all capacities are equal (namely, $C_i = C/n_c$, where n_c denotes the number of channels in the net), to be referred to as *identical* capacity assignment.

[2] This corresponds to continually searching the list.

number (the preferred path) is the leftmost entry in this position, and the rightmost number is the least desired acceptable path.

We now define a *fixed routing procedure* as the special case of alternate routing in which each list contains exactly one entry. Clearly, this procedure defines a unique path through the net for any origin-destination pair.

We consider four topologically different nets, each consisting of 13 nodes. The values of γ, μ, and C are left unspecified and are adjusted in the actual simulation so as to impose a range of traffic loads ρ on the net. Furthermore, we define three relative traffic matrices τ_1, τ_2, and τ_3 as shown below and on pages 110 and 111.

The first net we define is the *diamond* net. Its topological structure, its incidence matrix I_D, and the matrix R_{D1} (which defines its complete alternate routing procedure) are shown in Fig. 7.1. In all diagrams of network topology, the absence of an arrow on a channel indicates two independent one-way channels. The fixed routing procedure for the diamond net defined by the matrix R_{D4} is shown in Fig. 7.2. As may be noted, the entries in R_{D4} correspond to the leftmost entries in R_{D1}.

$\tau_1 =$ (Node of origination)

		Destination node											
	1	2	3	4	5	6	7	8	9	10	11	12	13
1	0	1	4	1	1	9	36	9	1	1	4	1	100
2	1	0	4	1	1	4	1	1	1	1	1	0	1
3	4	4	0	4	9	4	4	4	9	1	1	1	4
4	1	1	4	0	1	1	1	4	1	0	1	1	1
5	1	1	9	1	0	4	36	4	100	1	9	1	1
6	9	4	4	1	4	0	4	1	4	4	4	1	9
7	36	1	4	1	36	4	0	4	36	1	4	1	36
8	9	1	4	4	4	1	4	0	4	1	4	4	9
9	1	1	9	1	100	4	36	4	0	1	9	1	1
10	1	1	1	0	1	4	1	1	1	0	4	1	1
11	4	1	1	1	9	4	4	4	9	4	0	4	4
12	1	0	1	1	1	1	1	4	1	1	4	0	1
13	100	1	4	1	1	9	36	9	1	1	4	1	0

Also in Fig. 7.2 are shown the matrices R_{D2} and R_{D3} which describe other forms of alternate routing. In particular, R_{D3} allows the same alternate routing for messages destined for node 13 as does R_{D1}; however, messages destined for any other node in the net follow the fixed routing procedure of R_{D4}. Further, R_{D2} allows the full alternate routing only for messages in node i which are destined for node j, where the permitted values of the pair ij are (13, 1), (9, 5), (5, 9), and (1, 13). For other values of the pair ij, fixed routing is used. The motivation for defining these four routing procedures is to introduce varying degrees of alternate routing in the net. Specifically, if we consider the traffic matrix τ_1, we see that the heavy traffic flow is between the four origin-destination pairs listed for R_{D2}. Thus, for τ_1, R_{D4} represents no alternate routing (this is true for any τ_i), R_{D3} represents a small degree of alternate routing, R_{D2} represents a greater degree of alternate routing, and R_{D1} represents the largest degree of alternate routing that we consider.

We next present the *K-connected* net, defined in Fig. 7.3. This net corresponds to the *K*-connected net studied in Chap. 6 with $K = 4$.

Destination node

$\tau_2 =$ Node of origination

	1	2	3	4	5	6	7	8	9	10	11	12	13
1	0	1	4	1	1	9	36	9	1	1	4	1	9
2	1	0	4	1	1	4	1	1	1	1	1	0	1
3	4	4	0	4	9	4	4	4	9	1	1	1	4
4	1	1	4	0	1	1	1	4	1	0	1	1	1
5	1	1	9	1	0	4	36	4	9	1	9	1	1
6	9	4	4	1	4	0	4	1	4	4	4	1	9
7	100	1	100	1	36	100	0	100	36	1	100	1	100
8	9	1	4	4	4	1	4	0	4	1	4	4	9
9	1	1	9	1	9	4	36	4	0	1	9	1	1
10	1	1	1	0	1	4	1	1	1	0	4	1	1
11	4	1	1	1	9	4	4	4	9	4	0	4	4
12	1	0	1	1	1	1	1	4	1	1	4	0	1
13	9	1	4	1	1	9	36	9	1	1	4	1	0

Note that only a fixed routing procedure R_K has been defined for this net.

Another net of interest is the *fully connected* net, defined in Fig. 7.4. The incidence matrix I_F for this net contains an entry of unity in each off-diagonal position and an entry of zero in each main diagonal position. We define, for the fully connected net, an alternate routing matrix $R_F(n)$ for $n = 1, 2, \ldots, 6$. The value of n indicates the number of alternatives to be included in each ij position of the matrix. Figure 7.5 displays $R_F(6)$; $R_F(n)$ for $n < 6$ is merely the same as $R_F(6)$ with the last $6 - n$ entries deleted from each position. Clearly, $R_F(1)$ is the fixed routing procedure for this net. In order to fit Fig. 7.5 on one page, we have found it convenient to use a slightly different method for recording each list. Specifically, we have broken the list of six numbers into two lines of three numbers each, in such a way that the entry (1,2,3,4,5,6) becomes $\begin{pmatrix} 1,2,3 \\ 4,5,6 \end{pmatrix}$. Note that the fully connected net is identical with the K-connected net where K equals N.

Destination node

$\tau_3 = $

		1	2	3	4	5	6	7	8	9	10	11	12	13
	1	0	1	1	1	1	1	1	1	1	1	1	1	1
	2	1	0	1	1	1	1	1	1	1	1	1	1	1
	3	1	1	0	1	1	1	1	1	1	1	1	1	1
	4	1	1	1	0	1	1	1	1	1	1	1	1	1
	5	1	1	1	1	0	1	1	1	1	1	1	1	1
Node of origination	6	1	1	1	1	1	0	1	1	1	1	1	1	1
	7	1	1	1	1	1	1	0	1	1	1	1	1	1
	8	1	1	1	1	1	1	1	0	1	1	1	1	1
	9	1	1	1	1	1	1	1	1	0	1	1	1	1
	10	1	1	1	1	1	1	1	1	1	0	1	1	1
	11	1	1	1	1	1	1	1	1	1	1	0	1	1
	12	1	1	1	1	1	1	1	1	1	1	1	0	1
	13	1	1	1	1	1	1	1	1	1	1	1	1	0

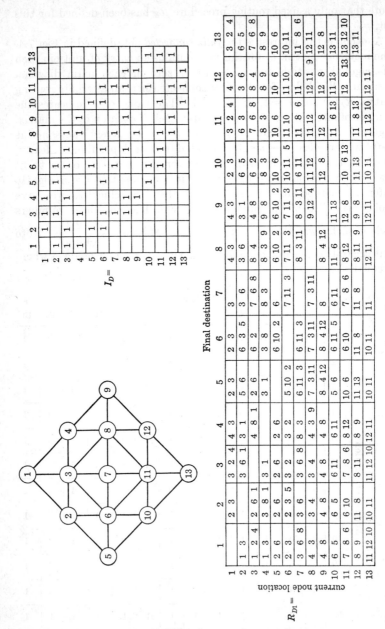

Fig. 7.1. *Diamond net with incidence matrix and complete alternate routing procedure.*

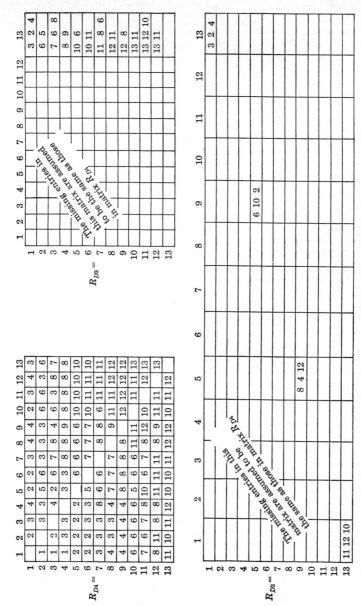

Fig. 7.2. *Other routine procedures for the diamond net.*

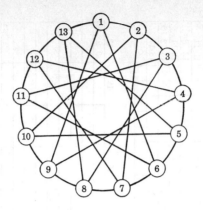

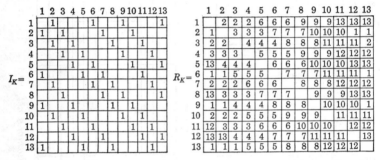

$I_K =$

	1	2	3	4	5	6	7	8	9	10	11	12	13
1		1				1			1				1
2	1		1				1			1			
3		1		1				1			1		
4			1		1				1			1	
5				1		1				1			1
6	1				1		1				1		
7		1				1		1				1	
8			1				1		1				1
9	1			1				1		1			
10		1			1				1		1		
11			1			1				1		1	
12				1			1				1		1
13	1				1			1				1	

$R_K =$

	1	2	3	4	5	6	7	8	9	10	11	12	13
1		2	2	2	6	6	6	9	9	9	13	13	13
2	1		3	3	3	7	7	7	10	10	10	1	1
3	2	2		4	4	4	8	8	8	11	11	11	2
4	3	3	3		5	5	5	9	9	9	12	12	12
5	13	4	4	4		6	6	6	10	10	10	13	13
6	1	1	5	5	5		7	7	7	11	11	11	1
7	2	2	2	6	6	6		8	8	8	12	12	12
8	13	3	3	3	7	7	7		9	9	9	13	13
9	1	1	4	4	4	8	8	8		10	10	10	1
10	2	2	2	5	5	5	9	9	9		11	11	11
11	12	3	3	3	6	6	6	10	10	10		12	12
12	13	13	4	4	4	7	7	7	11	11	11		13
13	1	1	1	5	5	5	8	8	8	12	12	12	

Fig. 7.3. *K-connected net* $(K = 4)$.

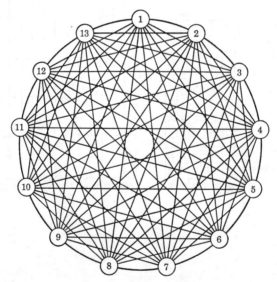

Fig. 7.4. *Fully connected net.*

$$R_F(n) =$$

1	2	3	4	5	6	7	8	9	10	11	12	13
	2 3 4 5 6 7	3 4 5 6 7 8	4 5 6 7 8 9	5 6 7 8 9 10	6 7 8 9 10 11	7 8 9 10 11 12	8 9 10 11 12 13	9 10 11 12 13 2	10 11 12 13 2 3	11 12 13 2 3 4	12 13 2 3 4 5	13 2 3 4 5 6
1 3 4 5 6 7		3 4 5 6 7 8	4 5 6 7 8 9	5 6 7 8 9 10	6 7 8 9 10 11	7 8 9 10 11 12	8 9 10 11 12 13	9 10 11 12 13 1	10 11 12 13 1 3	11 12 13 1 3 4	12 13 1 3 4 5	13 1 3 4 5 6
1 2 4 5 6 7	2 4 5 6 7 8		4 5 6 7 8 9	5 6 7 8 9 10	6 7 8 9 10 11	7 8 9 10 11 12	8 9 10 11 12 13	9 10 11 12 13 1	10 11 12 13 1 2	11 12 13 1 2 4	12 13 1 2 4 5	13 1 2 4 5 6
1 2 3 5 6 7	2 3 5 6 7 8	3 5 6 7 8 9		5 6 7 8 9 10	6 7 8 9 10 11	7 8 9 10 11 12	8 9 10 11 12 13	9 10 11 12 13 1	10 11 12 13 1 2	11 12 13 1 2 3	12 13 1 2 3 5	13 1 2 3 5 6
1 2 3 4 6 7	2 3 4 6 7 8	3 4 6 7 8 9	4 6 7 8 9 10		6 7 8 9 10 11	7 8 9 10 11 12	8 9 10 11 12 13	9 10 11 12 13 1	10 11 12 13 1 2	11 12 13 1 2 3	12 13 1 2 3 4	13 1 2 3 4 6
1 2 3 4 5 7	2 3 4 5 7 8	3 4 5 7 8 9	4 5 7 8 9 10	5 7 8 9 10 11		7 8 9 10 11 12	8 9 10 11 12 13	9 10 11 12 13 1	10 11 12 13 1 2	11 12 13 1 2 3	12 13 1 2 3 4	13 1 2 3 4 5
1 2 3 4 5 6	2 3 4 5 6 8	3 4 5 6 8 9	4 5 6 8 9 10	5 6 8 9 10 11	6 8 9 10 11 12		8 9 10 11 12 13	9 10 11 12 13 1	10 11 12 13 1 2	11 12 13 1 2 3	12 13 1 2 3 4	13 1 2 3 4 5
1 2 3 4 5 6	2 3 4 5 6 7	3 4 5 6 7 9	4 5 6 7 9 10	5 6 7 9 10 11	6 7 9 10 11 12	7 9 10 11 12 13		9 10 11 12 13 1	10 11 12 13 1 2	11 12 13 1 2 3	12 13 1 2 3 4	13 1 2 3 4 5
1 2 3 4 5 6	2 3 4 5 6 7	3 4 5 6 7 8	4 5 6 7 8 10	5 6 7 8 10 11	6 7 8 10 11 12	7 8 10 11 12 13	8 10 11 12 13 1		10 11 12 13 1 2	11 12 13 1 2 3	12 13 1 2 3 4	13 1 2 3 4 5
1 2 3 4 5 6	2 3 4 5 6 7	3 4 5 6 7 8	4 5 6 7 8 9	5 6 7 8 9 11	6 7 8 9 11 12	7 8 9 11 12 13	8 9 11 12 13 1	9 11 12 13 1 2		11 12 13 1 2 3	12 13 1 2 3 4	13 1 2 3 4 5
1 2 3 4 5 6	2 3 4 5 6 7	3 4 5 6 7 8	4 5 6 7 8 9	5 6 7 8 9 10	6 7 8 9 10 12	7 8 9 10 12 13	8 9 10 12 13 1	9 10 12 13 1 2	10 12 13 1 2 3		12 13 1 2 3 4	13 1 2 3 4 5
1 2 3 4 5 6	2 3 4 5 6 7	3 4 5 6 7 8	4 5 6 7 8 9	5 6 7 8 9 10	6 7 8 9 10 11	7 8 9 10 11 13	8 9 10 11 13 1	9 10 11 13 1 2	10 11 13 1 2 3	11 13 1 2 3 4		13 1 2 3 4 5
1 2 3 4 5 6	2 3 4 5 6 7	3 4 5 6 7 8	4 5 6 7 8 9	5 6 7 8 9 10	6 7 8 9 10 11	7 8 9 10 11 12	8 9 10 11 12 1	9 10 11 12 1 2	10 11 12 1 2 3	11 12 1 2 3 4	12 1 2 3 4 5	

Fig. 7.5. *Alternate routing for the fully connected net.*

115

Lastly, we present the *star* net, defined in Fig. 7.6. Note that for this net we define only a fixed routing procedure.

At this point it is convenient to comment on the sample size used in the simulation experiments. Clearly, we desire a sample sufficiently large so as to obtain an accurate measure of the recorded statistics. We quote a result due to Morse [14] for a single exponential channel facility which gives an approximate expression for D, the "relaxation" time to go from a mean square deviation away from the average queue length back to within a fraction $1/e$ (where e is the base of natural logarithms) of this deviation:

$$\mu CD \cong \frac{2\rho}{(1-\rho)^2}$$

where we have altered his notation to correspond to that used in our Appendix A. The interesting behavior to observe is that the relaxation time D depends upon the inverse square of $1 - \rho$; this exposes a great sensitivity to ρ as $\rho \to 1$. In order to illustrate a similar dependence in a network of queues, we show in Fig. 7.7 values of μCT (obtained from simulation of the diamond net with traffic matrix τ_1) as a function of M, the number of messages delivered to their destination; the param-

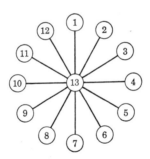

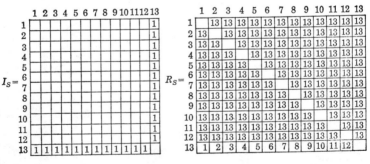

Fig. 7.6. *Star net.*

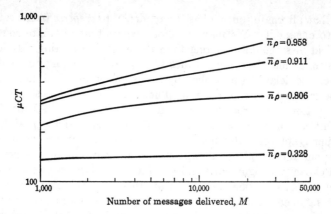

Fig. 7.7. *Effect of sample size on the average message delay (for the diamond net and traffic matrix τ_1).*

eter for this figure is $\bar{n}\rho$. As expected, we observe that the size of M required (to give a significant estimate of μCT) is strongly dependent on the value of $\bar{n}\rho$. In all simulations described in this monograph, the procedure used to determine when a sufficient number of messages had been processed was to observe μCT at various values of M; when these observed values leveled off sufficiently, the simulation was halted.

7.2 Simulation Results

Some results of the simulation programs have already been described in previous chapters. In Sec. 3.5 we discussed the use of the program in analyzing the dependent character of the traffic as it leaves a node. The simulation was used to investigate experimentally the validity of the conservation law (see Chap. 5) for nets. In Sec. 6.5 we saw how the program was used to generate experimental results for a K-connected net with fixed routing, so that these results might be compared with derived results for random routing procedures. We now proceed to describe a set of experimental simulations for the purpose of comparing the message delays for different capacity-assignment rules, different routing procedures, and different network topologies. We note here that the simulations were run *without* the use of the independence assumption.

The results of Theorems 4.2 and 4.4 and of Eq. (4.2) indicate that the average message delay in a queueing system is minimized when the traffic is concentrated into as few channels as possible. We note, however, that alternate routing has the tendency to *disperse* traffic rather than to cluster it. We may therefore expect that alternate routing will yield message delays larger than those of fixed routing. Indeed, this is so, as may be observed in Fig. 7.8a to d. The figure displays the message delay T (multiplied by μC for purposes of normalization) for traffic flow through the diamond net. Figure 7.8a to c shows μCT as a function of the routing procedure (R_{D1}, R_{D2}, R_{D3}, R_{D4}) for three different values of $\rho = \gamma/\mu C$ with traffic matrix τ_1. Figure 7.8d shows a similar graph for traffic matrix τ_2. Three curves are shown in each figure: one for the identical capacity assignment, one for the proportional capacity assignment,[1] and one for the square root assignment.

The first observation we make from these curves is that the square root assignment results in a considerably smaller message delay than does either of the other two assignments; in fact, it is preferable over a number of other capacity assignments that were tried experimentally. Not only was the square root assignment superior in the case of fixed routing procedures (as predicted by Theorem 4.5), but also in the case of various degrees of alternate routing.

The second observation is that the introduction of increasing degrees of alternate routing (plotted on an arbitrary horizontal scale) resulted in a deterioration of performance (when the square root assignment was used). For the identical and proportional capacity assignments, the introduction of alternate routing improved the performance at low values of $\gamma/\mu C$. One notes in Fig. 7.8a to c that the alternate routing described by the matrix R_{D1} resulted in a convergence of the three curves. This is not a general result (as evidenced by Fig. 7.8d), but is due both to the fact that the alternate routing procedure for τ_1 did not result in important path-length changes for the messages and to the fact that in the case of full alternate routing, the traffic tends to distribute itself in proportion to the capacities of the channels, thus resulting effectively in a system in which traffic and capacity are proportional; consequently, this routing procedure defeats the square root assignment of capacity. This convergence was not observed in the curves of Fig. 7.8d, since the introduction of alternate routing produced a significant change in the message path length.

Figure 7.9 shows similar plots for the fully connected net with traffic matrix τ_1. Once again, we observe that the square root assignment is superior to both the proportional and the identical capacity assignments, and that fixed routing is superior to alternate routing for a square root assignment rule. Further, it is clear that any alternate

---

[1] This assignment of capacity in proportion to the traffic carried results in a constant utilization factor for each channel in the net and is a reasonable assignment to consider on intuitive grounds.

path in this fully connected net must at least double the path length as compared to fixed routing, and so we do not see the convergence at heavy alternate routing here that we saw in Fig. 7.8a to c.

Whereas the introduction of alternate routing caused longer message delays (for the optimum channel capacity assignment), we find that the simulation results expose the ability of alternate routing procedures to *adapt* the flow of traffic so as to match the network topology. For example, Fig. 7.8b illustrates the improved performance due to alternate routing in the case of a poor capacity assignment (the identical

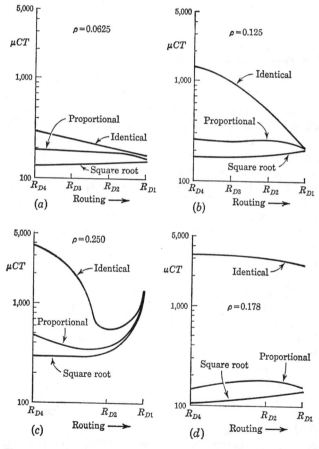

Fig. 7.8. *Effect of routing procedure and channel capacity assignment on the average message delay for the diamond net. (a), (b), and (c) Traffic matrix τ_1; (d) traffic matrix τ_2.*

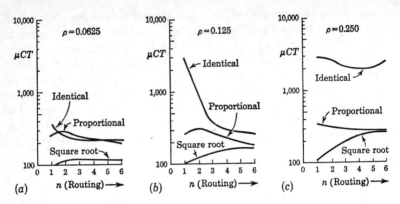

Fig. 7.9. *Effect of routing procedure and channel capacity assignment on the average message delay for the fully connected net (traffic matrix τ_1).*

capacity assignment). This adaptive behavior of alternate routing procedures has considerable significance in the realistic design and operation of a communication net. Specifically, it is generally true that the actual traffic matrix is not known precisely at the time the network is being designed; indeed, even if the traffic matrix were known, it is probable that the entries γ_{jk} in this matrix would be time-varying (i.e., different traffic loads would exist at different hours of the day, different days of the week, different seasons of the year, etc.). In the face of this uncertainty and/or variation, it becomes impossible to calculate the optimum channel capacity assignment from Eq. (4.7), since the numbers λ_i (which are calculable from the γ_{jk} under a fixed routing procedure) are in doubt. One solution to this problem is to use some form of alternate routing which adapts the actual traffic flow to the network. Note, however, that a price must be paid for such flexibility, since fixed routing with the square root capacity assignment is itself superior to alternate routing (assuming we have known, time-invariant γ_{jk}), as may be seen in Figs. 7.8 and 7.9.

Having observed the effects of routing and channel capacity assignment, we now consider the effect of topology on the message delay. Once again, we refer to Theorems 4.2 and 4.4 and to Eq. (4.2), which suggest the clustering of traffic and capacity. The star-net topology achieves this clustering in a very efficient manner. This net has the property that as much traffic as possible is grouped into each channel, subject to the constraint that at least one channel must enter and leave each node in the net. This physical constraint arises because of the fact that the traffic matrix (i.e., the set of origins and destinations) is specified independently of the network design, and, in general, each node will serve as an origin for some traffic and as a destination for some other traffic. Note that the star net achieves exactly one channel leading into and out of each node, except for the central node. The cost in terms of additional path length is less than twice that of the fully connected net (which clearly represents the net with the minimum path length for each message).

The behavior of the star net is contrasted with those of the diamond, the K-connected, and the fully connected nets in Fig. 7.10a and b. This figure displays the experimentally obtained message delay for the four nets with traffic matrix τ_1; Fig. 7.10a is drawn for the square root capacity assignment, and Fig. 7.10b for the identical capacity assignment. We note first that the star net is superior in its operation over most of the investigated range for both capacity assignments. Furthermore, the square root assignment is clearly superior in all cases to the identical capacity assignment (this is not surprising, since the traffic in τ_1 is nonuniformly distributed). Figure 7.11a and b shows the same comparison for the uniform traffic matrix τ_3. Here again, the star net is best over the range tested. We note, however, that the square root assignment is not significantly better than the identical capacity assignment in this case. This points out the interesting fact that when all

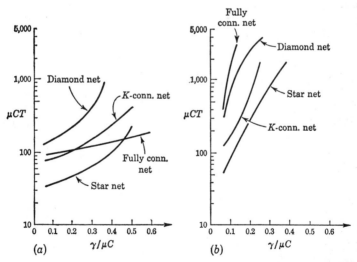

Fig. 7.10. *Effect of network topology on the average message delay for a fixed routing procedure and for the highly nonuniform traffic matrix τ_1. (a) Square root capacity assignment; (b) identical capacity assignment.*

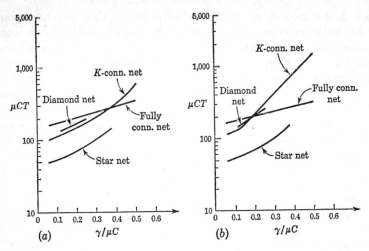

Fig. 7.11. *Effect of network topology on the average message delay for a fixed routing procedure and uniform traffic matrix* τ_3. *(a) Square root capacity assignment; (b) identical capacity assignment.*

λ_i are the same, it becomes clear from their definitions that all three capacity assignments yield the same results, namely, those of the identical capacity assignment.   Upon comparing Figs. 7.10a and 7.11a, we see that the square root assignment gives better results for the nonuniform traffic matrix τ_1 than for the uniform traffic matrix τ_3 (one should compare only the curves for the star and the fully connected nets here, since the average path length for the other two nets is a strong function of the traffic matrix).   The point to be made here is that the square root assignment increasingly gains advantage over the other forms of capacity allocation as the traffic pattern becomes more and more nonuniform; in fact, a highly nonuniform traffic pattern is precisely what Theorem 4.4 states will yield a minimum message delay.

Theorem 4.5 describes the optimum channel capacity assignment and gives the expression for the average message delay for a communication net with a fixed routing procedure subject to the constraint of constant total channel capacity.   The effects of the distribution of traffic λ_i and the average path length $\bar{n}$ on the average message delay may be seen in Eq. (4.18).   Specifically, we observe again that increased concentration of the traffic reduces the expression $\sum_i \sqrt{\lambda_i/\lambda}$ (where i ranges over the entire set of channels in the net).   Furthermore, we note that T grows without bound as $\rho \rightarrow 1/\bar{n}$, and so we desire to minimize $\bar{n}$ as well.   It is clear that the value of $\bar{n}$ is dependent upon the set of values

λ_i. In particular, for the star net (which has a maximally concentrated traffic pattern), $1 < \bar{n} < 2$. If we require a reduced $\bar{n}$, we must add channels to the star net, thus destroying some of the concentration of the traffic. In the limit as $\bar{n} \to 1$, we arrive at the fully connected net, which has the smallest possible $\bar{n}$ but also the most dispersed traffic pattern. The trade-off between $\bar{n}$ and traffic concentration depends heavily upon ρ. Thus we find that at low network loads nets similar in topology to the star net are optimum; as the network load increases, we obtain the optimum topology by reducing $\bar{n}$ (by adding additional channels); finally, as $\rho \to 1$, we require $\bar{n} = 1$, which results in the fully connected net. In all cases, we use the square root channel capacity assignment with a fixed routing procedure.[1] These conclusions were tested and verified by additional simulation experiments. In particular, we simulated just such a sequence of nets, starting with the star net, introducing additional channels, and finally arriving at the fully connected net. Figure 7.12 shows the results for the uniform

[1] It is interesting to note that the minimal-cost steady-flow problem (i.e., no storage in the system and a continuous constant flow) also leads to the fully connected topology with just enough capacity between each pair of nodes to satisfy the required flow. This solution, as in our present case, assumes no notion of distance measure between the nodes.

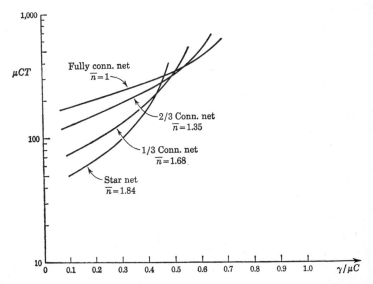

Fig. 7.12. *Effect of adding channels to the star net with fixed routing procedure, square root channel capacity assignment, and uniform traffic matrix τ_3.*

traffic matrix τ_3 with fixed routing and a square root channel capacity assignment. The $\frac{1}{3}$- and $\frac{2}{3}$-connected nets refer to a star net to which have been added enough direct channels to achieve nets with one-third (or two-thirds) of the total number of possible connections. Clearly, at low values of $\rho = \gamma/\mu C$, the pole in Eq. (4.18) is of small consequence, and so the dominating term is $\left(\sum_i \sqrt{\lambda_i/\lambda}\right)^2$; in such a case the star net is optimum, since it minimizes this sum. As ρ increases, the pole takes on more importance, and so $\bar{n}$ must be decreased by adding links, which results in a sacrifice of concentrated traffic. Finally, as $\rho \to 1$, the denominator in Eq. (4.18) dominates, so we must minimize $\bar{n}$ with no regard to the concentration. The optimum behavior over the entire range of ρ is then the minimum envelope of the family of curves which correspond to this sequence of topologies. The optimum topology at any particular value of ρ may be found by observing the behavior of T in Eq. (4.18) as more direct channels are added; this function will pass through a minimum as the channels are introduced; the topology corresponding to this minimum represents the optimum topology (for the particular value of ρ considered). In the case of a nonuniform traffic matrix, we expect a slightly different behavior; this is illustrated in Fig. 7.13 for the highly nonuniform traffic matrix τ_1. Here we see

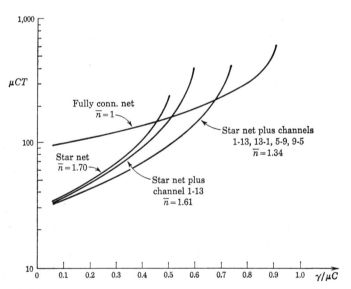

Fig. 7.13. *Effect of adding channels to the star net with fixed routing procedure, square root channel capacity assignment, and highly nonuniform traffic matrix τ_1.*

that even for small values of ρ the introduction of some direct links (between the nodes which carry the bulk of the traffic) is required. However, the addition of the remainder of the links only becomes necessary as ρ increases (as in the uniform traffic case).

7.3   Additional Theorems on Alternate Routing Procedures

In Sec. 7.2, we observed that the specific form of alternate routing procedure used gave rise to an increase in message delay under the square root assignment.   It was pointed out that this was not surprising in the light of our theoretical findings.   In that regard, we now present two theorems, one offering analytic proof of the superiority of fixed routing with additional constraints on the net, and the other analyzing a more sophisticated form of alternate routing.

Consider a net with a fixed total channel capacity fed by messages which have exponentially distributed interarrival times (i.e., Poisson arrivals) and lengths chosen independently from an exponential distribution each time the message enters a node (i.e., the independence assumption).   Also, consider any alternate routing procedure which results in internal network traffic which is Poisson.   For this situation, we state the following:

Theorem 7.1[1]

Consider any node (say node n_1) in the network from which there emanate two alternate paths[2] of the same length, both paths leading to the same node (say node n_2).   Then the message delay can always be reduced by omitting one of these paths as an alternative route for messages traveling from n_1 to n_2.

This theorem says, in essence, that a fixed route (i.e., a single path) is superior to alternate routing in the case in which the introduction of an alternative path does not affect the path length.   Of course, this result is restricted by the assumptions and only gives partial evidence of the advantages of fixed routing procedures.

The form of alternate routing procedure employed in the simulation experiments was extremely simple and unsophisticated.   We now discuss a refinement on the procedure for offering alternative routes to the message traffic.

We approach the problem from the point of view of a single node

[1] See Appendix F for proof of this theorem.
[2] By alternate paths we mean paths which are allowed by the routing procedure.

extracted from the net.   Consider such a node, with N channels emerging from it.   We assume that all these channels connect to either the same node or to a set of nodes, any one of which is an equally satisfactory next node to visit.   However, we assume that all the channels are of different capacities.   In fact, we choose to label the channels such that

$$C_1 > C_2 > C_3 > \cdots > C_N \qquad (7.1)$$

The other assumptions about the system are as follows: arrivals are Poisson; message lengths are independently and exponentially distributed with mean length $1/\mu$; the queue discipline is first-come first-served.   When a channel becomes free, it is offered to the first message in the queue; if the first message does not want the channel, it is offered to the second message, etc.   If no messages want the free channel, it remains idle until possibly some new message entering the system desires it.

We now inquire as to the conditions under which a message should accept channel C_i.   Specifically, let n_i be that position in the queue at which a message should accept channel C_i (that is, any message in position less than n_i is not interested in accepting channel C_i).   The criterion used is such that a message will accept channel C_i if and only if the acceptance of this channel reduces the message's expected time spent in the system.   In all cases, we assume that no message has any knowledge about the exact length of any other message, including its own length.

Applying the above criterion to the system leads to the following:

Theorem 7.2[1]

In the system described above, a message should accept channel C_i if and only if its position n_i in the queue satisfies the following inequalities:

$$n_i - 1 < \frac{S_{i-1}}{C_i} - 1 \le n_i \qquad i = 2, 3, \ldots, N \qquad (7.2)$$

where
$$S_{i-1} = \sum_{j=1}^{i-1} C_j \qquad (7.3)$$

and
$$n_1 = 1$$

One observes that these rules, while not complicated, add an interesting degree of sophistication to the simpler alternate routing schemes considered earlier.

---

[1] See Appendix F for proof of this theorem.

7.4 Conclusions

We have considered message delay to be the measure of a net's performance, and we have inquired as to the effect on performance of variations in channel capacity assignment, routing procedure, and topological network structure.   In the simulation experiments described in this chapter, we have been able to demonstrate some interesting behavior in regard to these design parameters while holding fixed the total channel capacity assigned to the net.   Specifically, the average message delay is highly sensitive to the average path length $\bar{n}$ and to the distribution of λ_i (which can be thought of as representing the degree of concentration in the internal traffic).   These two interact with each other and may not be optimized independently.   The recognition of these two quantities as those underlying the behavior of the average message delay allows us to predict theoretically and to observe experimentally a number of useful characteristics of communication nets.   We list these below.

1. The square root channel capacity assignment as described in Eq. (4.7) results in superior performance as compared with a number of other channel capacity assignments.
2. The performance of a straightforward fixed routing procedure with a square root capacity assignment surpasses that of a simple alternate routing procedure.
3. The alternate routing procedure adapts the internal traffic flow to suit the capacity assignment (i.e., the bulk of the message traffic is routed to the high-capacity channels).   This effect is especially noticeable and important in the case of a poor capacity assignment, which may come about owing to uncertainty or variation in the applied message traffic.
4. A high degree of nonuniformity in the traffic matrix results in improved performance in the case of a square root channel capacity assignment (owing to a more concentrated traffic pattern).
5. The quantities essential to the determination of the average message delay are the average path length and the degree to which the traffic flow is concentrated.

It is interesting to note that the trade-off between the average path length and the concentration of traffic results in a sequence of optimum network topologies which range from nets similar to the star net to the fully connected net.   Indeed, the star net has the maximally concentrated traffic but an average path length close to 2, and the fully connected net has the minimum average path length ($\bar{n} = 1$) but the maximally dispersed traffic. These two extreme cases bound the range of optimum nets, and the optimum topology for a particular value of network load ρ lies somewhere between them.

Two theorems involving alternate routing procedures were proven. The first theorem gave partial evidence of the superiority of fixed routing procedures. The second theorem introduced and analyzed a new alternate routing procedure in which a message had the option of refusing a free channel of low capacity and waiting a little longer for a higher-capacity channel.

chapter **8**

Conclusion

8.1 Summary

The description of the experimental results obtained by simulation completes our present investigation of communication nets.   We found that the introduction of the independence assumption into our model of a communication net was necessary for the analytic treatment of the problem; as discussed in Chap. 3, the approximation due to this assumption is quite good.

Theorems 4.2 and 4.4 and Eq. (4.2) all indicate that message delay can be minimized by concentrating the traffic into a small number of high-capacity channels; the word small here is ideally to be interpreted as unity; however, certain physical constraints on the network often place a lower bound which exceeds unity on the number of channels. Theorem 4.5 describes the optimum channel capacity assignment and gives the expression for the average message delay for a communication net with a fixed routing procedure subject to the constraint of constant total channel capacity.   We recognize from this theorem that the average path length and the degree to which the internal traffic is concentrated play an essential role in determining the average delay of a message.

The study of priority disciplines led to two useful conclusions.   First, the delay-dependent priority structure provides the system designer with a number of degrees of freedom with which to manipulate the relative waiting times for each priority group.   Second, the conservation law (Theorem 5.4) allows one to draw a number of general conclusions about the average waiting times for a large class of priority structures.

The investigation of Chap. 6 pointed out that, relative to fixed routing procedures, the disadvantages of random routing procedures are: increased path lengths for messages, increased message delay at each node, and a reduction in the total average traffic that the network can accommodate.

The digital simulation program was used to study the behavior of a number of different nets, as described in Chap. 7. The program served a multiple purpose by allowing the verification of theoretically predicted results and the investigation of mathematically intractable problems; furthermore, a number of ideas for further study were obtained from the experiments. The major results of the simulation are listed at the end of Chap. 7.

The general problem discussed in this monograph has been the minimization of the average message delay with respect to the channel capacity assignment, the routing procedure, the priority discipline, and the network topology, subject to the constraint of a fixed total channel capacity assigned to the net. The problem statement may be expressed as

$$\underset{\begin{cases}\text{Capacity assignment,}\\ \text{routing procedure,}\\ \text{priority discipline,}\\ \text{topology}\end{cases}}{\text{Minimize}} \sum_{i=1}^{N} \frac{\lambda_i}{\gamma} T_i \qquad (8.1)$$

subject to
$$D = \sum_{i=1}^{N} d_i C_i \qquad (8.2)$$

where N is the number of channels in the net (N may vary as the topology varies). We note, from Eq. (C.1), that the above minimization of message delay is equivalent to minimizing the total average number of messages in the net; that is, $\lambda_i T_i$ is the average number of messages waiting for, or being transmitted on, the ith channel. Furthermore, when we compare the form of T [as expressed in Eq. (4.17)] with the conservation law (Theorem 5.4), we observe that all the results we have obtained for T apply equally well to any queue discipline which falls into the class defined by Theorem 5.4.

8.2 Suggestions for Further Investigations

The model of a communication net which we have chosen to investigate is clearly an idealization of real communication systems. It would be rather useful to relax a number of the restrictions, leaving a model still capable of analysis. For example, the desirability of a star-net topology would be highly questionable in a network which did not have perfectly reliable nodes and links. The assumption of noiseless communication channels has simplified our analysis; the removal of this restriction would probably lead to a system wherein the considerations were not vastly different from those in the noiseless case.[1] Introducing more general message traffic (such as multidestination messages and mixtures of data and direct traffic) would greatly extend the field of application of our communication network study. Along with the notion of message defection goes the possibility of finite storage capacity at each node; this generalization is also a line for future investigation. We restate the assumption of stationary Poisson-exponential statistics for the external message sources; more general distributions would be extremely interesting to study. One of the more difficult mathematical problems is that of solving for the statistics of the internal traffic flow within a communication net. An indication of the scope and source of the difficulty and some partial answers to the problem are given in Chap. 3; these complications led to the introduction of the independence assumption.

The alternate routing procedure used in the simulation experiments is of a rather elementary form; a more sophisticated procedure is described in Sec. 7.3 and bears further investigation.

An interesting generalization of the channel capacity assignment stated in Eq. (4.7) is given in Theorem 4.6. Specifically, the theorem solves for the optimum channel capacity assignment and for the resultant average message delay for a communication net subject to the following constraint [which replaces the constraint expressed in Eq. (4.4)]:

$$D = \sum_{i=1}^{N} d_i C_i \qquad (8.3)$$

where C_i is the channel capacity of the ith transmission channel, and d_i is a function independent of the capacity C_i which reflects the cost, say in dollars, of supplying one unit of channel capacity to the ith channel. The quantity D represents the total number of dollars that are available to spend in supplying the N-channel system with the set of capacities C_i. The actual values of the set d_i depend upon the par-

---

[1] In the case of a noisy channel, some form of coding must be used so that transmission with an arbitrarily small probability of error is achieved. The effect of such coding is to add additional delays to the message transmission. Since delay is the measure of performance in our analysis it appears that the extension to noisy channels is straightforward.

ticular communication net involved; for example, d_i might be chosen to represent the length of the ith channel. It would be interesting to see exactly what one could say about the topological design of a net with this new set of constraints; perhaps the topologies involved would consist of many small groups of nodes, each group connected together in a manner not unlike that described in Chap. 7, and the set of groups then connected together in a similar topology, and so on (i.e., a hierarchy of topologies, the basic element of which is chosen as one of the members in the sequence which ranges from the star net to the fully connected net).

As mentioned above, the notion of unreliable nodes and links is one that would bear consideration. Indeed, a most interesting problem statement which incorporates the extension to unreliability as well as to the more general cost function described by Eq. (8.3) may be expressed as

$$\text{Minimize} \quad AT + BU$$
$$\begin{Bmatrix} \text{Capacity assignment,} \\ \text{routing procedure,} \\ \text{priority discipline,} \\ \text{topology} \end{Bmatrix}$$

subject to the fixed-cost constraint expressed in Eq. (8.3)

where A and B represent one's dislike for a unit of message delay T and unreliability U, respectively. Of course, the function U must consist of an appropriately defined measure of the unreliability.

Most of the suggested extensions listed above bring one face to face with mathematically intractable problems. It seems appropriate to suggest that rather than perform an exact analysis on these extended models, one should approach the problem with the intent of obtaining approximate or asymptotic analytic results. Furthermore, the instrument of verification for these results (and in some cases their only source) would be a communication network simulator such as that described in Appendix E.

A Review of Simple Queueing Systems[1]

A queue is a waiting line. Examples of queues occur constantly in our daily living: a line of cars in front of a toll booth, water backed up by a dam, customers at the check-out counter of a supermarket, messages awaiting transmission at a communication center, etc. In order to have a process in which queues can form, we require arrivals (of cars, water, people, messages) and a service facility which performs some operation on the arrivals (collecting the toll, releasing the water, checking out customers, transmitting messages).

Thus, we have a process involving *flow*. It becomes immediately obvious that if we hope to contend with this flow, we must ensure that the capacity of our servicing facility is sufficient to handle the *average* flow rate. If the flow is *steady*, then we have a straightforward problem.[2] If, however, the flow comes in spurts or in any other non-uniform fashion, then we can expect queues to form in front of the service facility, even if its capacity exceeds the average input flow rate. A queue forms when a higher than average flow occurs and saturates

---

[1] The material presented in this appendix is intended to acquaint the reader with some of the basic notions and results of queueing theory which have application in our study of communication networks. It might be well for those readers with a knowledge of elementary queueing theory to pass over this material.

[2] See, for example, the department-store escalator problem in Sec. 1.1.

the service facility; when the flow is less than average, the service facility may find itself idle for some period.

Queueing theory deals with the description and analysis of the effects of such a fluctuating flow. It is the purpose of this appendix to present certain well-known results for some simple queueing systems, so as to expose the characteristic behavior of queueing systems in general.

A.1   Birth and Death Equations: Exponential Assumptions

Before we involve ourselves with any particular queueing process, let us discuss a well-known general result for a class of birth and death processes. In particular, we assume that we have a population of units in a system in which we allow the population to increase and decrease according to some birth and death coefficients as follows. Let

$b_n \, dt$ = P$_r$ [birth of new unit occurs during time interval $(t, \, t + dt)$, given that there were n units in system at time t]

$d_n \, dt$ = P$_r$ [death of old unit occurs during time interval $(t, \, t + dt)$, given that there were n units in system at time t]

$o(dt)$ = P$_r$ [more than one event occurs in $(t, \, t + dt)$]*

where dt is the differential time interval. The assumption here is that the birth and death coefficients b_n and d_n are independent of time; clearly, they are functions of n, the number of units already in the system. It is also clear from the assumptions that the system can change only through transitions from states to their nearest neighbors (the system is said to be in *state n* when there are n units in the system). The boundary condition is that d_0 is identically zero.

Let us further define $P_n(t)$ as follows:

$P_n(t)$ = P$_r$ [finding n units in system at time t]

With these assumptions and definitions, one can write the Chapman-Kolmogorov equation, which leads to the following differential-difference equations[1]

$$\frac{dP_0(t)}{dt} = d_1 P_1(t) - b_0 P_0(t)$$

$$\frac{dP_n(t)}{dt} = d_{n+1} P_{n+1}(t) + b_{n-1} P_{n-1}(t) - (d_n + b_n) P_n(t) \qquad n \geq 1$$

These equations, known as the *forward birth and death equations*, are easily solved under the assumption that the birth and death coefficients

---

* The notation $o(dt)$ implies that $o(dt)/dt$ approaches zero as dt approaches zero.
[1] See Feller [18].

are independent of time. Specifically, this assumption leads to an exponential distribution of time spent in state n, with exponent $-(b_n + d_n)t$.

Let us now assume the existence of a limiting distribution for $P_n(t)$; that is,

$$\lim_{t \to \infty} P_n(t) = P_n$$

Furthermore, we consider only those cases[1] for which this limiting distribution is such that

$$\sum_{n=0}^{\infty} P_n = 1$$

We now set all time derivatives equal to zero in the forward equations and obtain a new set of time-independent equations whose solution is

$$P_n = P_0 \prod_{i=0}^{n-1} \left(\frac{b_i}{d_{i+1}} \right) \tag{A.1}$$

The constant P_0 can be determined from

$$P_0 = 1 - \sum_{n=1}^{\infty} P_n$$

Equation (A.1) is the important result, of which we shall make considerable use.

A large number of queueing processes can be described in terms of a birth and death process with suitably chosen coefficients b_n and d_n, as we shall see in the remainder of this appendix (and in the proof of some theorems in Chap. 6).

A.2 Single Exponential Channel[2]

A queueing process consists of a service facility and a population of units which arrive at this facility and require service. Since the facility may be engaged in servicing another unit when a new unit arrives, one must provide a "waiting room" (storage) in which arrivals

---

[1] See Feller [18].

[2] Throughout this appendix, as well as in most of this study, we shall be concerned with the particular case of messages arriving at a communication center. The service that these messages seek is to be transmitted out of this center to some other center via a communication channel. The server is considered to be busy whenever the communication channel is in use.

may queue up.   Once serviced, a unit leaves the system.   In order to describe the queueing process properly, one must specify the following:

1. The arrival statistics—in particular, the distribution of the inter-arrival times
2. The service statistics—in particular, the distribution of service times
3. The rules for forming and maintaining the queue (e.g., the maximum number that the storage facility can hold and the queue discipline)
4. The number of service facilities

One can add to the list many other specifications which describe various forms of queueing processes.   For a reasonably complete list, the reader is referred to Saaty [23].

For our purposes, let us describe an extremely simple and fundamental queueing process which exhibits the characteristic behavior of a large number of more complicated systems.   In particular, we assume the following:

1. The interarrival times are exponentially distributed with an average arrival rate of λ messages[1] per second.
2. The message lengths are exponentially distributed with an average length of $1/\mu$ bits per message; the channel capacity associated with the transmission facility is C bits/sec.
3. The queue discipline is first-come first-served with an infinite storage capability.
4. There is one channel available for transmission.

From these assumptions, one recognizes that the average time required to transmit a message over the channel is $1/\mu C$ seconds.   This effectively associates the variable aspect of service with the message rather than with the service facility.   This choice differs from the usual assumptions and introduces a new parameter C.   The reasoning behind this choice becomes obvious when one considers the physical system being discussed.

The assumptions above describe what is known as the *single exponential channel*.[2]   Its simplicity lies in the assumption of independent exponential distributions for interarrival times and message lengths.

---

[1] See footnote 2 on page 135.
[2] A huge literature has been devoted to the study of the single exponential channel; for example, see Morse [20] or Saaty [23].

As is well known, the exponential distribution is the only continuous distribution which exhibits a complete lack of memory (see Feller [18]).

The quantities of interest in any queueing process are numerous. We are not interested in time-dependent solutions, but rather we shall concentrate on the steady-state solutions, in which all transient effects have damped out. In particular, we are interested in the following:

1. $P_n = $ P$_r$ [n messages in system]
2. $E(n) = $ expected value of number of messages in system
3. $p(t)\, dt = $ P$_r$ [total time that message spends in system lies in interval $(t, t + dt)$]
4. $T = $ expected value of total time that message spends in system
5. $P(>n) = $ P$_r$ [more than n messages in system]
6. $p(>t) = $ P$_r$ [total time spent in system is greater than t]

Whenever we speak of the system, we include the queue as well as the service facility (the transmission channel in this case).

The solution for any of these quantities involves a new parameter ρ which may be defined as follows:

$$\rho = \frac{\lambda}{\mu C} \tag{A.2}$$

ρ is thus the ratio of the average input data rate to the maximum transmission rate. It turns out that ρ is also the fraction of time that the channel is busy, and so it is referred to as the *utilization factor*. As was stated in the introduction to this appendix, ρ must be less than 1 if the system is to be capable of handling the flow.[1] Even with ρ less than 1 we encounter a queueing effect, as will be obvious from the equations which follow shortly.

If we desire, we may consider the queueing process to be a birth and death process with coefficients

$$b_n = \lambda$$
and
$$d_n = \mu C$$

It is then clear from Eq. (A.1) that the solution for P_n is, for $\rho < 1$,

$$P_n = (1 - \rho)\rho^n \tag{A.3}$$

where we have evaluated P_0 so that the total probability sums to unity.

---

[1] When $\rho > 1$, no steady-state distributions exist (see Kendall [9]).

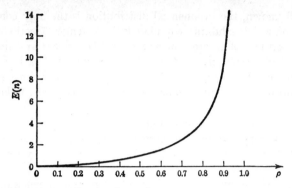

Fig. A.1. *Expected number of messages in the system for the single exponential channel.*

By perfectly straightforward methods,[1] one can solve for all quantities of interest listed above; they turn out to be

$$E(n) = \frac{\rho}{1 - \rho} \tag{A.4}$$

$$p(t) = \mu C(1 - \rho)e^{-(1-\rho)\mu Ct} \tag{A.5}$$

$$T = \frac{1}{\mu C(1 - \rho)} \tag{A.6}$$

$$P(>n) = \rho^{n+1} = e^{-(n+1)[\log C - \log \lambda/\mu]} \tag{A.7}$$

$$p(>t) = e^{-(1-\rho)\mu Ct} \tag{A.8}$$

In these results one can observe some characteristics of queueing processes in general. The most important observation to make is that the quantity $1 - \rho$ appears in the denominators of both expected-value expressions. Figure A.1 shows $E(n)$ as a function of ρ. The simple pole at $\rho = 1$ is common to a large number of queueing processes.[2] Physically, the expected number of messages in the system grows without bound as ρ approaches unity owing to the fact that the fluctuations in the input data rate cause the short-term average input data rate to vary above and below the long-term average; when the long-term average approaches the channel capacity (that is, ρ approaches unity), the

---

[1] Once again, the reader is referred to Morse [20] or Saaty [23].

[2] For example, in 1951, Kendall [9] showed that for a single-channel system identical with the one we have considered, except for an arbitrary distribution of message lengths (with mean lengths $1/\mu C$ and variance σ^2), the expected time that a message spends in the system is

$$T = \frac{1}{\mu C} + \frac{\rho^2 + \lambda^2\sigma^2}{2\lambda(1 - \rho)}$$

short-term variations often raise the input rate above the channel capacity and thus cause large increases in the queue length.

One also notes that in this simple process all distributions are either exponential or geometric. It is clear that the parameter ρ plays a central role in all the results; this is true of other queueing processes as well.

A.3 Multiple Exponential Channels

Instead of a single-channel facility, let us now consider a queueing process with N channels, each of capacity C/N. Thus the total capacity of the service facility is the same as in the single-channel case. In addition, the queue discipline is the same as before, except that when a message gets to the head of the queue it must take the first available channel. If a message enters the system when more than one channel is free, it chooses one from this set according to a uniform distribution. Otherwise, the assumptions about arrival and service statistics, etc., apply to this case (see Fig. 4.1).

The same quantities are of interest here, and once again we define ρ as the ratio of the average input data rate to the maximum transmission rate:

$$\rho = \frac{\lambda}{\mu C}$$

That is, when all channels are occupied, the transmission rate is C bits/sec.

Furthermore, we may consider this queueing process as a birth and death process with coefficients

$$b_n = \lambda$$

$$d_n = \begin{cases} \dfrac{n\mu C}{N} & n \leq N \\[2mm] \mu C & n \geq N \end{cases}$$

It is then clear from Eq. (A.1) that the solution for P_n is, for $\rho < 1$,

$$P_n = \begin{cases} \dfrac{P_0 \rho^n N^n}{n!} & n \leq N \\[2mm] \dfrac{P_0 \rho^n N^N}{N!} & n \geq N \end{cases} \tag{A.9}$$

where

$$P_0 = \left[\sum_{n=0}^{N-1} \frac{(N\rho)^n}{n!} + \frac{(N\rho)^N}{(1-\rho)N!} \right]^{-1} \tag{A.10}$$

By straightforward methods,[1] one can solve for the following quantities of interest (defined in Sec. A.2):

$$P(\geq N) = \frac{P_0(N\rho)^N}{(1-\rho)N!} \tag{A.11}$$

$$E(n) = \frac{\rho}{(1-\rho)}[N(1-\rho) + P(\geq N)] \tag{A.12}$$

$$p(t) = \left[\frac{N(1-\rho) - P(<N)}{N(1-\rho) - 1} e^{-(\mu C/N)t} \right. $$
$$\left. - \frac{N(1-\rho)P(\geq N)}{N(1-\rho) - 1} e^{-(1-\rho)\mu Ct} \right] \frac{\mu C}{N} \tag{A.13}$$

$$T = \frac{N}{\mu C} + \frac{P(\geq N)}{(1-\rho)\mu C} \tag{A.14}$$

$$p(>t) = \frac{N(1-\rho) - P(<N)}{N(1-\rho) - 1} e^{-(\mu C/N)t} - \frac{P(\geq N)}{N(1-\rho) - 1} e^{-(1-\rho)\mu Ct} \tag{A.15}$$

where $P(<N) = 1 - P(\geq N)$

Once again, one notes the central role played by the quantity ρ (the utilization factor) in all these expressions. In particular, as ρ approaches unity, the expected number of messages in the system grows without bound as in the single-channel case; this is the situation in most types of queueing processes.

A.4   The Output of a Queueing Process

An interesting theorem which describes the output distribution (i.e., the interdeparture-time distribution) from a multiple exponential channel system has been described by Burke [15]. We make use of this theorem in Chaps. 3 and 4. Following is a statement of the theorem (for the proof, see [15]):

Theorem A.1 (*Due to Burke*)

The steady-state output of a queue with N channels in parallel, with Poisson arrival statistics, and with lengths chosen independently from an exponential distribution is itself Poisson-distributed.

The reasons for presenting the equations for the multiple-channel facility should be clear from the area of interest of this research. The problem under consideration involves a network of communication centers. Each center has a number of communication channels connecting it to some other centers; thus each center begins to resemble the multiple-channel service facility. This relationship is discussed in Chaps. 3 and 4.

[1] See Morse [20] or Saaty [23].

appendix **B**

Theorems and Proofs for Chapter 4

B.1   Theorem 4.1 and Its Proof

Theorem 4.1[1]

Consider an N-channel service facility of total capacity C with Poisson arrivals (rate λ) and exponential message lengths (mean length $1/\mu$). Define

$$\rho = \frac{\lambda}{\mu C}$$

Then

$$\rho = 1 - \sum_{n=0}^{\infty} \frac{\bar{C}_n}{C} P_n.$$

provided

$$\rho < 1$$

where $\bar{C}_n$ = average unused capacity given n messages in system
$P_n = \mathrm{P_r}\,[n$ messages in system]

PROOF: The system considered satisfies the conditions of the birth and death process, examined in Appendix A, with

$$b_n = \lambda$$
$$d_n = \mu(C - \bar{C}_n)$$

[1] The statement of this theorem is somewhat abbreviated here; full details may be found in Chap. 4.

Therefore, by Eq. (A.1), we find that

$$P_n = \frac{P_0 \rho^n}{\prod_{i=1}^{n} (1 - r_i)} .$$

(B.1)

where

$$r_i = \frac{\bar{C}_i}{C}$$

Let us solve for P_0.

$$1 = \sum_{n=0}^{\infty} P_n = P_0 \left(1 + \sum_{n=1}^{\infty} R_n \rho^n\right)$$

where

$$R_n = \frac{1}{\prod_{i=1}^{n} (1 - r_i)}$$

Thus,

$$P_0 = \frac{1}{1 + \sum_{n=1}^{\infty} R_n \rho^n}$$

(B.2)

Now, let us form

$$x = 1 - \sum_{n=0}^{\infty} \frac{\bar{C}_n}{C} P_n$$

according to the statement of the theorem. Noting that $\bar{C}_0 = C$ by construction and using Eqs. (B.1) and (B.2), we obtain

$$x = 1 - \frac{1 + \sum_{n=1}^{\infty} r_n R_n \rho^n}{1 + \sum_{n=1}^{\infty} R_n \rho^n}$$

$$= \frac{\sum_{n=1}^{\infty} R_n \rho^n (1 - r_n)}{1 + \sum_{n=1}^{\infty} R_n \rho^n}$$

and so $x = \rho$, which proves the theorem.

B.2   Theorem 4.2 and Its Proof

Theorem 4.2

The value of N which minimizes T for the system shown in Fig. 4.1, for all $0 \le \rho < 1$, is

$$N = 1$$

PROOF: From Eqs. (A.10), (A.11), and (A.14),

$$T = \frac{N}{\mu C}\left[1 + \frac{1/N(1-\rho)}{S_N(1-\rho)+1}\right]$$ (B.3)

where

$$S_N = \sum_{n=0}^{N-1} \frac{(N\rho)^{n-N}N!}{n!} > 0$$

Now

$$S_N = \sum_{n=0}^{N-1} \rho^{n-N} \frac{N}{N}\frac{N-1}{N}\cdots\frac{n+1}{N}$$

Therefore

$$S_N \le \sum_{n=0}^{N-1} \rho^{n-N} = \frac{\rho^{-N}-1}{1-\rho}$$

giving

$$0 < S_N \le \frac{\rho^{-N}-1}{1-\rho} \qquad \text{for } N \ge 1$$ (B.4)

Now, Eq. (B.3) yields

$$T = \frac{1}{\mu C(1-\rho)} \qquad \text{for } N = 1$$

Therefore, it is sufficient to show that

$$T > \frac{1}{\mu C(1-\rho)} \qquad \text{for all } N > 1 \text{ and } 0 \le \rho < 1$$

Using Eq. (B.4), we get, for Eq. (B.3),

$$T \ge \frac{N}{\mu C}\left[1 + \frac{\rho^N}{N(1-\rho)}\right]$$

$$= \frac{N(1-\rho)+\rho^N}{\mu C(1-\rho)}$$

Letting

$$\alpha = 1 - \rho$$

we see that

$$N(1-\rho)+\rho^N = N\alpha + (1-\alpha)^N$$
$$\ge N\alpha + 1 - N\alpha$$
$$= 1$$

Thus,

$$T \ge \frac{1}{\mu C(1-\rho)}$$

for all N; the only case for which the equality holds is $N = 1$. Note that the equality would also hold for $\alpha = 0$, but this implies that $\rho = 1$, which we do not permit. Thus

$$T > \frac{1}{\mu C(1-\rho)} \qquad \text{for } N > 1 \text{ and } 0 \le \rho < 1$$

which completes the proof.

B.3   Theorems 4.3, 4.5, and 4.6 and Their Proofs

For purposes of proof only, we consider a more general case (similar to Theorem 4.6) in which we allow an arbitrary set of μ_i (whereas Theorem 4.6, in its original form, requires that $\mu_i = \mu$ for all i in order to obtain a proper physical interpretation). After proving this more general case, we observe that Theorem 4.3 is the special case wherein $D = C$ and $d_i = 1$ for all i; Theorem 4.5 is the special case wherein $D = C$, $d_i = 1$, and $\mu_i = \mu$ for all i; and Theorem 4.6 is the special case wherein $\mu_i = \mu$ for all i. Thus, we propose to prove the three theorems simultaneously.

Accordingly, we first state this more general theorem below (for purposes of proof only):

Theorem

Consider a communication net with N channels, such as is described for Theorem 4.6, in which we now allow an arbitrary set of μ_i. The assignment of channel capacity C_i to the ith channel which minimizes T subject to the constraint

$$D = \sum_{i=1}^{N} C_i d_i \tag{B.5}$$

is

$$C_i = \frac{\lambda_i}{\mu_i} + \frac{D_e}{d_i} \frac{\sqrt{\lambda_i d_i / \mu_i}}{\sum_{j=1}^{N} \sqrt{\lambda_j d_j / \mu_j}} \tag{B.6}$$

With this optimum assignment,

$$T_i = \frac{\sum_{j=1}^{N} \sqrt{\lambda_j d_j / \mu_j}}{D_e \sqrt{\lambda_i \mu_i / d_i}} \tag{B.7}$$

and

$$T = \frac{\bar{n} \left(\sum_{i=1}^{N} \sqrt{\frac{\lambda_i d_i}{\lambda \mu_i}} \right)^2}{D_e} \tag{B.8}$$

provided

$$D_e > 0$$

where

$$D_e = D - \sum_{j=1}^{N} \frac{\lambda_j d_j}{\mu_j}$$

and

$$\bar{n} = \frac{\lambda}{\gamma} \tag{B.9}$$

PROOF: We recognize, from Eq. (4.17), that

$$T = \sum_{i=1}^{N} \frac{\lambda_i}{\gamma} T_i$$

As discussed in Sec. 4.4, the independence assumption allows us to write T_i as [see also Eq. (A.6)]

$$T_i = \frac{1}{\mu_i C_i - \lambda_i}$$

With these expressions, we now use the method of Lagrange multipliers[1] as follows: Let

$$G = T + \alpha \left(\sum_{i=1}^{N} C_i d_i - D \right)$$

where α is some undetermined constant multiplier.   Differentiating G with respect to C_i and setting this result equal to zero, we obtain

$$0 = - \frac{\lambda_i}{\gamma} \frac{\mu_i}{(\mu_i C_i - \lambda_i)^2} + \alpha d_i$$

or
$$C_i = \frac{\lambda_i}{\mu_i} + \frac{1}{\sqrt{\alpha \gamma}} \sqrt{\frac{\lambda_i}{\mu_i d_i}} \tag{B.10}$$

Multiplying Eq. (B.10) by d_i and summing on i, we find

$$\sum_{i=1}^{N} C_i d_i = D = \sum_{i=1}^{N} \frac{\lambda_i d_i}{\mu_i} + \frac{1}{\sqrt{\alpha \gamma}} \sum_{i=1}^{N} \sqrt{\frac{\lambda_i d_i}{\mu_i}}$$

from which we obtain

$$\frac{1}{\sqrt{\alpha \gamma}} = \frac{D - \sum\limits_{i=1}^{N} (\lambda_i d_i / \mu_i)}{\sum\limits_{i=1}^{N} \sqrt{\lambda_i d_i / \mu_i}} \tag{B.11}$$

Substituting Eq. (B.11) into Eq. (B.10), we arrive at

$$C_i = \frac{\lambda_i}{\mu_i} + \frac{D_e}{d_i} \frac{\sqrt{\lambda_i d_i / \mu_i}}{\sum\limits_{j=1}^{N} \sqrt{\lambda_j d_j / \mu_j}}$$

which establishes Eq. (B.6).

[1] See Hildebrand [53].

If we now substitute Eq. (B.6) into our expression for T_i, we obtain

$$T_i = \frac{1}{\mu_i \left[\dfrac{\lambda_i}{\mu_i} + \dfrac{D_e}{d_i} \dfrac{\sqrt{\lambda_i d_i/\mu_i}}{\sum\limits_{j=1}^{N} \sqrt{\lambda_j d_j/\mu_j}} \right] - \lambda_i}$$

or

$$T_i = \frac{\sum\limits_{j=1}^{N} \sqrt{\lambda_j d_j/\mu_j}}{D_e \sqrt{\lambda_i \mu_i/d_i}}$$

This establishes Eq. (B.7).

Furthermore, substituting Eq. (B.7) into our expression for T gives us

$$T = \sum_{i=1}^{N} \frac{\lambda_i}{\gamma} T_i = \sum_{i=1}^{N} \frac{\lambda_i}{\gamma} \frac{\sum\limits_{j=1}^{N} \sqrt{\lambda_j d_j/\mu_j}}{D_e \sqrt{\lambda_i \mu_i/d_i}}$$

or

$$T = \frac{\bar{n} \left(\sum\limits_{i=1}^{N} \sqrt{\lambda_i d_i/\lambda \mu_i} \right)^2}{D_e}$$

This establishes Eq. (B.8).

In order to show that $\bar{n} = \lambda/\gamma$, we observe that $\bar{n}$ may be written as

$$\bar{n} = \sum_{j,k} \frac{\gamma_{jk}}{\gamma} n_{jk} \tag{B.12}$$

where n_{jk} is the path length for the origin-destination pair jk. Now, we recognize that λ_i is the sum of all γ_{jk} for which the jk route includes channel i. If we consider λ (i.e., the sum of the λ_i by definition), we observe that it is composed of the numbers γ_{jk}, each added in n_{jk} times. That is,

$$\lambda = \sum_{i} \lambda_i = \sum_{j,k} \gamma_{jk} n_{jk} \tag{B.13}$$

If we now substitute Eq. (B.13) into Eq. (B.12), we obtain

$$\bar{n} = \frac{\lambda}{\gamma}$$

which establishes Eq. (B.9) and completes the proof of the theorem stated above.

We observe that setting $\mu_i = \mu$ for all i (in the theorem above) establishes Theorem 4.6. For Theorems 4.3 and 4.5, we take $D = C$

and $d_i = 1$ for all i; we observe that under these conditions

$$D_e = C - \sum_{j=1}^{N} \frac{\lambda_j}{\mu_j}$$

$$= C - \lambda \sum_{j=1}^{N} \frac{\lambda_j}{\lambda} \frac{1}{\mu_j}$$

From the definitions of $1/\mu$ and ρ given in Sec. 1.5, and from the expression for $\bar{n}$ in Eq. (B.9), we find that

$$D_e = C - \frac{\lambda}{\mu}$$

or

$$D_e = C(1 - \bar{n}\rho)$$

Using this expression for D_e in the theorem as stated in this appendix and noting that $\bar{n} = 1$ for the conditions of Theorem 4.3, we prove Theorem 4.3. By setting $\mu_i = \mu$ and using this expression for D_e, we establish Theorem 4.5.

B.4 Theorem 4.4 and Its Proof

Theorem 4.4

The distribution of λ_i which minimizes T in Theorem 4.3 subject to the constraints expressed by Eqs. (4.6) and (4.13) is, for $\mu_i = \mu$,

$$\lambda_i = \begin{cases} \lambda - \sum_{j=2}^{N} k_j & i = 1 \\ k_i & i = 2, 3, \ldots, N \end{cases}$$

PROOF: We note that minimizing Eq. (4.12), with $\mu_i = \mu$, is equivalent to minimizing the sum S:

$$S = \sum_{i=1}^{N} \sqrt{\lambda_i}$$

We prove the theorem by showing that if any λ_i (say λ_j) is of the form

$$\lambda_j = k_j + \alpha_j$$

where $\alpha_j > 0$, then S will not be increased by the new assignment

$$\lambda'_j = k_j$$
$$\lambda'_1 = \lambda_1 + \alpha_j$$

where $j \geq 2$. Specifically, we wish to show that for

$$S' = \sqrt{\lambda_1'} + \sqrt{\lambda_j'} + \sum_{\substack{i=2 \\ i \neq j}}^{N} \sqrt{\lambda_i}$$

the following holds:

$$S' \leq S$$

where the prime indicates a single change of the form described above. We must therefore show that

$$\sqrt{\lambda_1'} + \sqrt{\lambda_j'} \leq \sqrt{\lambda_1} + \sqrt{\lambda_j}$$

or

$$\sqrt{\lambda_1 + \alpha_j} + \sqrt{k_j} \leq \sqrt{\lambda_1} + \sqrt{k_j + \alpha_j}$$

Since both sides of the last inequality are positive, we may square each side to obtain, as our condition,

$$\lambda_1 + \alpha_j + k_j + 2\sqrt{(\lambda_1 + \alpha_j)k_j} \leq \lambda_1 + k_j + \alpha_j + 2\sqrt{\lambda_1(k_j + \alpha_j)}$$

Once again, owing to the positiveness of these quantities, we may obtain, by cancellation and squaring,

$$k_j \leq \lambda_1 \tag{B.14}$$

Equation (B.14) is the necessary and sufficient condition for our theorem to hold. Now, by Eq. (4.14), we have

$$k_j \leq k_1$$

But, by definition,

$$\lambda_1 = k_1 + \alpha_1$$

where $\alpha_1 \geq 0$. Thus, we obtain

$$k_j \leq k_1 \leq \lambda_1$$

which establishes Eq. (B.14) and proves the theorem.

We note that if $\alpha_1 = 0$ and $k_j = k_1$, then $S' = S$; otherwise, $S' < S$. Thus, if more than one $\lambda_j = k_j + \alpha_j$, where $\alpha_j > 0$, then surely $S' < S$ after all these changes (indicated by the primed variables above) are made successively.

Theorems and Proofs
for Chapter 5

We make extensive use of a well-known result in queueing theory throughout this appendix. The result was conjectured by many researchers, and recently a formal proof of its validity was given by Little [17]. Roughly stated, it says that the expected number $E(n)$ of units in a queueing system which has reached equilibrium is equal to the product of input rate λ of these units to the system and the expected value τ of the time spent by these units in the system; i.e.,

$$E(n) = \lambda\tau \qquad (C.1)$$

Certain weak restrictions are placed upon the queueing process, but these need not concern us, since all the systems with which we deal satisfy these conditions.

The definition of system in this equality is left unspecified, and so we may choose to define it as the queue itself, in which case we use the notation $\tau = W$; or we may choose to define it as the system which includes both the queue and the service facility, in which case we use the notation $\tau = T$. In addition, we may choose to separate the units in the system into a set of subgroups, in which case the equality above holds for each subgroup separately [i.e., labeling the pth subgroup with the subscript p, we have $E(n_p) = \lambda_p \tau_p$, where τ_p may take the form W_p or T_p, depending upon the choice of definition for the system].

149

Most of the proofs of the priority-queueing results in this appendix are based upon arguments concerning expected values. In such cases, the random variables involved are usually suppressed. In order to offer one example of a complete derivation involving the random variables, the proof of the conservation law is carried out in its entirety. This complete procedure leads naturally to Eq. (C.5), involving only expected values, which could have been written immediately if one had originally argued on an expected-value basis. Most of the other proofs are carried out using expected values only, since the random-variable arguments for all proofs are so similar.

Many of the theorems proven in this Appendix involve the assumption that service times (i.e., message lengths) are chosen independently from some exponential distribution. It is well known[1] that an exponential distribution is the only continuous distribution with a Markovian character; i.e., it is memoryless. By this we mean that if, for a nonnegative random variable x,

$$\mathrm{P_r}[x \geq t] = e^{-\mu t}$$

then, for $t \geq t_0 \geq 0$,

$$\mathrm{P_r}[x \geq t | x \geq t_0] = e^{-\mu(t-t_0)} \tag{C.2}$$

This property is easily checked by carrying out the calculation, and we make extensive use of it.

We find it convenient to derive the theorems in this Appendix by first proving the Conservation Law (and its related corollaries) and then proceeding in the order given in Chap. 5.

C.1   The Conservation Law (Theorem 5.4) and Its Proof

Theorem 5.4

For any queue discipline and any fixed arrival and service-time distributions which fall into the class defined in Sec. 5.2,

$$\sum_{p=1}^{P} \rho_p W_p = \begin{cases} \dfrac{\rho}{1 - \rho} V_1 & 0 \leq \rho < 1 \\ \infty & \rho \geq 1 \end{cases}$$

where

$$V_1 = \tfrac{1}{2} \sum_{p=1}^{P} \lambda_p E(t_p{}^2)$$

and

$E(t_p{}^2)$ = second moment of service-time distribution for group p

[1] See, for example, Feller [18, Sec. XVII.6].

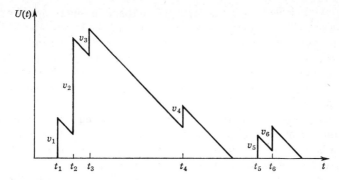

Fig. C.1. *Total unfinished work $U(t)$ in the system.*

PROOF: Let us define $U(t)$ as the total unfinished work present in the system at time t. In particular, $U(t)$ represents the time it would take to empty the system of all messages present at time t, if no new messages (units)[1] were allowed to enter the system after time t. A typical section of $U(t)$ might be similar to the graph shown in Fig. C.1.

The instants t_i are the times of arrival (independent and Poisson) of new units to the system, each unit having its service time v_i chosen independently from some distribution. The $U(t)$ function decreases at a steady rate of 1 sec/sec as long as $U(t)$ is positive; it jumps by v_i at the times t_i, and, once having reached zero, it remains there until the arrival of the next unit. Now, it is clear that the following limit is well defined (and exists whenever $\rho < 1$ for the system under consideration):

$$\bar{W} = \lim_{T \to \infty} \frac{1}{T} \int_0^T U(t) \, dt$$

Thus $\bar{W}$ is defined as the time-averaged value of $U(t)$.

Let us restrict the class of queueing systems that we consider to those which satisfy the conditions set forth in Sec. 5.2. Under these restrictions, it is clear that no matter what discipline is used (priority, preemption, or what have you), as long as the same set of t_i and v_i are involved, the function $U(t)$ will be the same. It is further obvious that no matter which $U(t)$ function turns up, as long as the same statistics are used for the t_i and v_i, the expected value $\bar{W}$ of the unfinished work will (with probability one) be the same.

One recognizes that the expected value of the waiting time (in queue only) for a unit in a strict first-come first-served discipline is just $\bar{W}$

---

[1] The terms *message* and *unit* are used interchangeably in this appendix.

(i.e., the waiting time is exactly equal to the unfinished work in a first-come first-served system). We make use of the independence of $\bar{W}$ of the particular discipline used by first deriving an expression for the expected value of the waiting time (in queue) for an arbitrary queue discipline [see Eq. (C.5)] and then evaluating $\bar{W}$ in this equation for the first-come first-served system. We consider first the case $0 \leq \rho < 1$.

Accordingly, let the random variable w represent the waiting time (in queue) of an arbitrary unit (the tagged unit). In addition, let there be n_p type p units present in the queue upon the arrival of the tagged unit; also, let t_{ip} represent the time spent in service by the ith unit ($i = 1, 2, \ldots, n_p$) of type p. Finally, let t_0 be the time required to complete service on the unit found in service upon the arrival of the tagged unit. Thus, w may be written as

$$w = t_0 + \sum_{p=1}^{P} \sum_{i=1}^{n_p} t_{ip} \qquad \text{(C.3)}$$

We have separated the units in the system into P classes. This is done in anticipation of applying the result of this derivation to priority systems, etc., which have P classes of units. Now, w, t_0, t_{ip}, and n_p are all random variables. Let us next form the expected value[1] on both sides of Eq. (C.3):

$$\bar{W} = V_1 + \sum_{p=1}^{P} \sum_{n_p=0}^{\infty} r(n_p) \sum_{i=1}^{n_p} E(t_{ip}) \qquad \text{(C.4)}$$

where, clearly,[2] $E(t_0) \equiv V_1$, and $r(n_p)$ is the probability that n_p type p units are present in the queue upon the arrival of the tagged unit. We define

$$E(t_{ip}) = \frac{1}{\mu_p}$$

where all service times for type p units are chosen independently from the same distribution (not necessarily exponential) whose mean is

---

[1] Note that Eq. (C.3) is capable of yielding more relationships of the type of Eq. (5.18). These may be obtained by first raising Eq. (C.3) to the nth power ($n = 2, 3, \ldots$) and then taking expected values.

[2] Equation (5.19), which gives an explicit expression for V_1, has been derived by a number of authors; for example, a simple derivation may be found in Saaty [23, Sec. 11-2.1a].

$1/\mu_p$. Thus, Eq. (C.4) becomes

$$\bar{W} = V_1 + \sum_{p=1}^{P} \frac{1}{\mu_p} \sum_{n_p=0}^{\infty} n_p r(n_p)$$

Now, from Eq. (C.1), we recognize that

$$\sum_{n_p=0}^{\infty} n_p r(n_p) \equiv E(n_p) = \lambda_p W_p$$

where λ_p and W_p are consistent with their earlier definitions. Thus we arrive at the following form for $\bar{W}$:

$$\bar{W} = V_1 + \sum_{p=1}^{P} \rho_p W_p \tag{C.5}$$

Let us now evaluate $\bar{W}$ for a strict first-come first-served discipline with Poisson input traffic (since $\bar{W}$ is invariant to a change in queue discipline within our class); this implies that all (average) waiting times W_p are equal, and in particular $W_p = \bar{W}$ for all p. Thus, we convert Eq. (C.5) to

$$\bar{W} = \frac{V_1}{1 - \rho} \tag{C.6}$$

Substituting the value of $\bar{W}$, as given by Eq. (C.6), into Eq. (C.5), we obtain

$$\sum_{p=1}^{P} \rho_p W_p = \frac{\rho}{1 - \rho} V_1$$

which establishes Eq. (5.18) for $0 \leq \rho < 1$.

For the case $\rho \geq 1$ we need only recognize that the input traffic rate exceeds[1] the service rate, in which case we see immediately that at least one of the W_p (where $\rho_p > 0$) grows without bound. Of course, in such a case we have no steady-state solution. This completes the proof of the conservation law.

It is convenient to digress at this point in order to illustrate a simple method for establishing the result

$$V_1 = \sum_{p=1}^{P} \frac{\rho_p}{\mu_p}$$

---

[1] For $\rho \to 1$, we note that $[\rho/(1 - \rho)]V_1$ approaches ∞. The limiting case for $\rho = 1$ is discussed fully by Lindley [11].

for exponentially distributed service times. Let us define T_p as equal to $W_p + 1/\mu_p$. Also define $1/\mu'_p$ as the expected value of the additional service time required by a unit of type p, given that this unit was still in the system at an arbitrary instant of time. Accordingly, the expected value of the unfinished work is

$$\bar{W} = \sum_{p=1}^{P} \frac{\lambda_p T_p}{\mu'_p} \qquad (C.7)$$

where, once again, we have used Eq. (C.1). Taking advantage of the memoryless property of exponential distributions as discussed previously [see Eq. (C.2)], we come to the conclusion that

$$\frac{1}{\mu'_p} = \frac{1}{\mu_p}$$

Using this and the substitution $T_p = W_p + 1/\mu_p$ in Eq. (C.7), we find that

$$\bar{W} = \sum_{p=1}^{P} \frac{\rho_p}{\mu_p} + \sum_{p=1}^{P} \rho_p W_p$$

Comparing this with Eq. (C.5), we conclude that, for exponential service times,

$$V_1 = \sum_{p=1}^{P} \frac{\rho_p}{\mu_p}$$

The origin of the definition of W_0 ($=V_1$) in Sec. 5.1 is now clear.

C.2 Corollaries to the Conservation Law and Their Proofs

There exist priority disciplines for which $\rho \geq 1$, and for which a subclass of the priority groups obtain a bounded steady-state solution for W_p. In particular, this is true for the fixed-priority systems studied in Sec. 5.1. If we consider the fixed-priority systems with $0 \leq \rho$, we expect that some of the W_p may grow without bound; let us label this set with the indices $p = 1, 2, \ldots, j - 1$. For $p = j, j + 1, \ldots,$ P, we expect[1] bounded W_p. The conservation law holds, of course, but we wonder what constraints on the waiting time may exist for those groups with $p \geq j$. We express these constraints in the following two corollaries.

---

[1] Once again, the reader is referred to Phipps [47].

Corollary 1

For $0 \leq \rho$ and a fixed-priority discipline with no preemption, under restrictions 1, 2, and 4 in Sec. 5.2,

$$\sum_{p=j}^{P} \rho_p W_p = \frac{s_j}{1 - s_j} (V_j + V_j') \qquad \text{(C.8)}$$

where

$$V_j = \tfrac{1}{2} \sum_{p=j}^{P} \lambda_p E(t_p^2) \qquad \text{(C.9)}$$

$$V_j' = \frac{f}{2} \lambda_{j-1} E(t_{j-1}^2) \qquad \text{(C.10)}$$

$$j = \text{smallest positive integer such that } \sum_{p=j}^{P} \rho_p < 1$$

$$s_j = \sum_{p=j}^{P} \rho_p \qquad \text{(C.11)}$$

$$\text{and} \qquad f = \begin{cases} 0 & \rho < 1 \\ \dfrac{1 - s_j}{\rho_{j-1}} & \rho \geq 1 \end{cases} \qquad \text{(C.12)}$$

PROOF: Let

$U_j = E$ (remaining work in system to be done on classes j, $j + 1, \ldots, P$)

$V_j = E$ (remaining work in service facility to be done on classes $j, j + 1, \ldots, P$)

$V_j' = E$ (remaining work in service facility to be done on class $j - 1$)

$f\rho_{j-1} = $ fraction of time that a unit from class $j - 1$ is in service facility

Now, considering a nonpreemptive fixed-priority system and following the same type of argument that led to Eq. (C.5), we see that

$$U_j = V_j + \sum_{p=j}^{P} \rho_p W_p \qquad \text{(C.13)}$$

We now wish to evaluate U_j, which is invariant to any priority structure within the classes $j, j + 1, \ldots, P$ for the same reasons that $\bar{W}$ was invariant in the case $\rho < 1$. Taking advantage of this, we evaluate U_j by setting up the following simplified priority structure. Let all units in the classes $j, j + 1, \ldots, P$ follow a first-come first-served discipline among themselves; however, they join the queue in front of all units of priority $j - 1$ or less. In such a case, for $p \geq j$, we can write

$$W_p = U_j + V'_j$$

That is, the time spent in the queue will be U_j (owing to the first-come first-served structure) plus the time required to complete the unit found in service if that unit is from class $j - 1$ (since if the unit in service is from class $\geq j$, this additional time is accounted for in U_j). Thus Eq. (C.13) becomes

$$U_j = V_j + \sum_{p=j}^{P} \rho_p U_j + \sum_{p=j}^{P} \rho_p V'_j \qquad (C.14)$$

and so

$$U_j = \frac{V_j + s_j V'_j}{1 - s_j} \qquad (C.15)$$

where

$$s_j = \sum_{p=j}^{P} \rho_p$$

Equating Eqs. (C.13) and (C.15), we get

$$\sum_{p=j}^{P} \rho_p W_p + V_j = \frac{V_j + s_j V'_j}{1 - s_j}$$

which yields

$$\sum_{p=j}^{P} \rho_p W_p = \frac{s_j}{1 - s_j} (V_j + V'_j)$$

Owing to the Poisson input statistics, we may apply the result of Cobham [48] and Phipps [47];[1] i.e.,

$$V_j = \tfrac{1}{2} \int_0^\infty t^2 \sum_{p=j}^{P} \lambda_p \, dF_p(t) = \tfrac{1}{2} \sum_{p=j}^{P} \lambda_p E(t_p^2)$$

and

$$V'_j = \frac{f}{2} \int_0^\infty t^2 \lambda_{j-1} \, dF_{j-1}(t) = \frac{f}{2} \lambda_{j-1} E(t_{j-1}^2)$$

Also, we notice that $f\rho_{j-1}$, the fraction of the time during which type $j - 1$ units utilize the service facility, may be calculated as

$$f\rho_{j-1} = 1 - s_j \qquad \text{for } \rho \geq 1$$

and so

$$f = \frac{1 - s_j}{\rho_{j-1}} \qquad \text{for } \rho \geq 1 \text{ (or } j > 1)$$

For completeness, we define

$$f = 0 \qquad \text{for } \rho < 1 \ (j = 1)$$

---

[1] The cumulative distribution function for the service time of the pth priority group is denoted by $F_p(t)$.

With these substitutions, we note that for the case $\rho < 1$ we obtain the result given in Eq. (5.18), which, of course, we must. This completes the proof of Corollary 1.

Corollary 2

For a fixed-priority discipline with a preemptive-resume rule and exponential service-time distributions under restrictions 1, 2, and 4 in Sec. 5.2,

$$\sum_{p=j}^{P} \rho_p W_p = \frac{s_j}{1 - s_j} V_j \qquad (C.16)$$

PROOF: We now show how an equation similar to Eq. (C.8) may be obtained for a preemptive-resume situation with exponentially distributed service times. Clearly, Eq. (C.13) still holds. In order to evaluate U_j, we now use the same trick as in the nonpreemption case (i.e., we form all priority classes into two groups—the first group consisting of classes $j, j+1, \ldots, P$, and all other classes being in the second group), except that we allow members of the first group to preempt units of the second group. Then we see that, for $p \geq j$,

$$W_p = U_j$$

and so Eq. (C.13) becomes

$$U_j = V_j + \sum_{p=j}^{P} \rho_p U_j$$

or

$$U_j = \frac{V_j}{1 - s_j} \qquad (C.17)$$

Substituting Eq. (C.17) into Eq. (C.13) yields

$$\frac{V_j}{1 - s_j} = V_j + \sum_{p=j}^{P} \rho_p W_p$$

or

$$\sum_{p=j}^{P} \rho_p W_p = \frac{s_j}{1 - s_j} V_j$$

Once again, V_j is as given previously. Note also that, for $j = 1$, Eq. (C.16) reduces to Eq. (5.18). This completes the proof of Corollary 2.

We note here that in the case of exponentially distributed service times we obtain

$$V_j = \sum_{p=j}^{P} \frac{\rho_p}{\mu_p}$$

and so we recognize that $V_1 = W_0$ as defined in Sec. 5.1.

C.3   Theorem 5.1 and Its Proof

Theorem 5.1

For a fixed-priority system with preemption and $0 \leq \rho$,

$$
W_p = \begin{cases} \dfrac{\dfrac{\rho_p}{\mu_p} + \displaystyle\sum_{i=p+1}^{P} \rho_i \left(\dfrac{1}{\mu_p} + \dfrac{1}{\mu_i}\right) + \displaystyle\sum_{i=p+1}^{P} \rho_i W_i}{1 - \displaystyle\sum_{i=p}^{P} \rho_i} & p \geq j \\[2em] \infty & p < j \end{cases}
$$

or

$$
W_p = \begin{cases} \dfrac{\dfrac{s_j}{1 - s_j} \displaystyle\sum_{i=j}^{P} \dfrac{\rho_i}{\mu_i} + \dfrac{\rho_p}{\mu_p} + \displaystyle\sum_{i=p+1}^{P} \rho_i \left(\dfrac{1}{\mu_p} + \dfrac{1}{\mu_i}\right) - \displaystyle\sum_{i=j}^{p-1} \rho_i W_i}{1 - \displaystyle\sum_{i=p+1}^{P} \rho_i} & p \geq j \\[2em] \infty & p < j \end{cases}
$$

where j is as defined in Cobham's result, and

$$
s_j = \sum_{i=j}^{P} \rho_i
$$

PROOF: We argue on the basis of expected values in proving this theorem.   Let

T_i = expected time that a message from priority class i spends in
     system (queue plus service time)

The expected time T_p for a unit (say the tagged unit) from priority class p is composed of three terms: its expected time in service, the expected time to service units (with as high or higher priority) present at its time of arrival, and the expected time to service units (with higher priority) which enter the system while the tagged unit is still in the system.[1]   The expected number of messages from the ith priority group which are present at the time of arrival of the tagged unit is, by Eq. (C.1), $\lambda_i T_i$.   The expected number of messages from the ith priority group which arrive during the time that the tagged unit remains in the system is, by consideration of Poisson arrival statistics, $\lambda_i T_p$.

---

[1] Recall that units within the same priority class are served in a first-come first-served fashion.

We collect these statements in the following equation, where we assume $0 \leq \rho < 1$:

$$T_p = \frac{1}{\mu_p} + \sum_{i=p}^{P} \frac{1}{\mu_i} \lambda_i T_i + \sum_{i=p+1}^{P} \frac{1}{\mu_i} \lambda_i T_p \tag{C.18}$$

Now, clearly,

$$W_i = T_i - \frac{1}{\mu_i}$$

and so Eq. (C.18) becomes

$$W_p = \sum_{i=p}^{P} \rho_i \left(W_i + \frac{1}{\mu_i} \right) + \sum_{i=p+1}^{P} \rho_i \left(W_p + \frac{1}{\mu_p} \right)$$

Solving this, we obtain, for $0 \leq \rho < 1$,

$$W_p = \frac{\rho_p/\mu_p + \sum\limits_{i=p+1}^{P} \rho_i(1/\mu_p + 1/\mu_i) + \sum\limits_{i=p+1}^{P} \rho_i W_i}{1 - \sum\limits_{i=p}^{P} \rho_i}$$

Now, for $\rho \geq 1$ we expect stable behavior only for $p \geq j$, where j is the smallest positive integer such that $\sum\limits_{i=j}^{P} \rho_i < 1$ (see Phipps [47]). The derivation above will now apply only for $p \geq j$; in the case $p < j$, we have $W_p = \infty$. Noting this, we see that we have proven Eq. (5.9) of Theorem 5.1.

We now establish Eq. (5.10). Observe that

$$\sum_{i=p+1}^{P} \rho_i W_i = \sum_{i=j}^{P} \rho_i W_i - \sum_{i=j}^{p} \rho_i W_i$$

Now, using Corollary 2, we get

$$\sum_{i=p+1}^{P} \rho_i W_i = \frac{s_j}{1 - s_j} \sum_{i=j}^{P} \frac{\rho_i}{\mu_i} - \sum_{i=j}^{p} \rho_i W_i$$

Using this last expression in Eq. (5.9) yields, for $p \geq j$,

$$W_p = \frac{[s_j/(1 - s_j)] \sum\limits_{i=j}^{P} \rho_i/\mu_i + \rho_p/\mu_p + \sum\limits_{i=p+1}^{P} \rho_i(1/\mu_p + 1/\mu_i) - \sum\limits_{i=j}^{p-1} \rho_i W_i}{1 - \sum\limits_{i=p+1}^{P} \rho_i}$$

which establishes Eq. (5.10) of Theorem 5.1.

C.4 Theorem 5.2 and Its Proof

Theorem 5.2

For the delay-dependent priority system with no preemption and $0 \le \rho < 1$,

$$W_p = \frac{W_0/(1 - \rho) - \sum_{i=1}^{p-1} \rho_i W_i (1 - b_i/b_p)}{1 - \sum_{i=p+1}^{P} \rho_i (1 - b_p/b_i)}$$

or

$$W_p = \frac{W_0}{1 - \rho} \frac{1}{D_p} \left[1 + \sum_{j=1}^{p-1} \sum_{0 < i_1 < i_2 < \cdots < i_j < p} F_{i_1}(i_2) F_{i_2}(i_3) \cdots F_{i_j}(p) \right]$$

where

$$D_p = 1 - \sum_{i=p+1}^{P} \rho_i \left(1 - \frac{b_p}{b_i} \right)$$

and

$$F_k(n) = - \frac{\rho_k}{D_k} \left(1 - \frac{b_k}{b_n} \right)$$

PROOF: We now establish Eq. (5.12), in which the priority for a unit which is assigned a parameter b_p $(0 \le b_1 \le \cdots \le b_P)$ is calculated at time t as follows:

$$q_p(t) = (t - T) b_p$$

where T is the time of arrival of the unit. We refer to such a unit as being of type p.

Consider the arrival of a type p unit, the tagged unit. Upon its arrival, the expected number $E(n_i)$ of type i units present in the queue is, by Eq. (C.1),

$$E(n_i) = \lambda_i W_i$$

Let f_{ip} represent the expected fraction of these type i units which receives service before the tagged unit. As usual, W_p will represent the expected value of the time that the tagged unit spends in the queue. We know, by assumption, that the expected number $E(m_i)$ of type i units which arrive during the time interval W_p is

$$E(m_i) = \lambda_i W_p$$

That this is so is obvious from the definition of λ_i as the average number of type i arrivals per second and the independence of arrival times. Let g_{ip} represent the expected fraction of these $E(m_i)$ type i units which receives service before the tagged unit. Further, let us define W_0 as we have in the past, namely, as the expected value of the time required to complete service on the unit found in service upon entry.

With these observations and definitions, we are able to write down a set of P simultaneous equations, one for each value of p, as follows:

$$W_p = W_0 + \sum_{i=1}^{P} \frac{\lambda_i W_i}{\mu_i} f_{ip} + \sum_{i=1}^{P} \frac{\lambda_i W_p}{\mu_i} g_{ip} \qquad (C.19)$$

The typical term in these sums is of the following form: the expected number of type i units which get service before the tagged unit multiplied by the quantity $1/\mu_i$ (which is the expected value of the service time for a type i unit).

Now, from the definitions of f_{ip} and g_{ip} as well as the imposed queue discipline, we note that

$$f_{ip} = 1 \qquad \text{for all } i \geq p$$
and
$$g_{ip} = 0 \qquad \text{for all } i \leq p$$

Using this information and solving for W_p in Eq. (C.19), we obtain

$$W_p = \frac{W_0 + \sum_{i=p}^{P} \rho_i W_i + \sum_{i=1}^{p-1} \rho_i W_i f_{ip}}{1 - \sum_{i=p+1}^{P} \rho_i g_{ip}} \qquad (C.20)$$

Let us now derive an expression for g_{ip}. Once again consider the arrival of a type p unit, the tagged unit, at time 0. Since W_p is its expected waiting time, the expected value of its attained priority at the expected time it is accepted for service is $b_p W_p$, as shown in Fig. C.2. In deriving g_{ip}, we must calculate the number of type i units

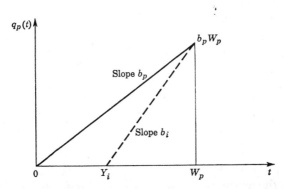

Fig. C.2. *Diagram of priority $q_p(t)$ for obtaining g_{ip}.*

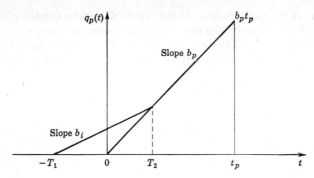

Fig. C.3.   *Diagram of priority $q_p(t)$ for obtaining f_{ip}.*

arriving, on the average, after time zero and reaching a priority of at least $b_p W_p$ before time W_p.   It is obvious from the figure that type i units which arrive in the time interval $(0, Y_i)$ will satisfy these conditions.   Thus, let us calculate the value of Y_i.   Clearly,

$$b_p W_p = b_i (W_p - Y_i)$$

and so

$$Y_i = W_p \left(1 - \frac{b_p}{b_i} \right)$$

Therefore, with an input rate of λ_i for the type i units, we find that

$$g_{ip} E(m_i) = \lambda_i Y_i$$

and so

$$g_{ip} \lambda_i W_p = \lambda_i W_p \left(1 - \frac{b_p}{b_i} \right)$$

giving

$$g_{ip} = 1 - \frac{b_p}{b_i} \qquad \text{for all } i > p$$

We now prove that $f_{ip} = b_i / b_p$ for $i \leq p$.   Consider that a type p unit, the tagged unit, arrives at time $t = 0$ and spends a total time t_p in the queue.   Its attained priority at the time of its acceptance into the service facility will be $b_p t_p$, as shown in Fig. C.3.

Upon its arrival, the tagged unit finds n_i type i units already in the queue.   Let us consider one such type i unit, as shown in the figure, which arrived at $t = -T_1$.   In deriving f_{ip}, we must calculate the number of type i units arriving before $t = 0$ and obtaining service before the tagged unit.   It is obvious from the figure that a type i unit which arrives at time $-T_1$ $(T_1 > 0)$ and which waits in the queue a time $w_i(T_1)$ such that $T_1 \leq w_i(T_1) \leq T_1 + T_2$ will satisfy these conditions.   Obviously, if $w_i(T_1) > T_1 + T_2$, the i type unit will be of lower priority than the tagged unit and will therefore fail to meet the conditions stipulated above.   Note that T_2 may exceed t_p, but this does not violate our conditions, since in that case the type i unit must surely be serviced before the tagged unit is serviced.

Let us first solve for T_2. Clearly,

$$b_p T_2 = b_i (T_1 + T_2)$$

and so

$$T_2 = \frac{b_i}{b_p - b_i} T_1$$

or

$$T_1 + T_2 = \frac{b_p}{b_p - b_i} T_1$$

It is clear that the expected number $E(n_i) f_{ip}$ of type i units in the queue at $t = 0$ which obtain service before the tagged unit can be expressed as

$$E(n_i) f_{ip} = \int_0^\infty \lambda_i P_r \left[t \leq w_i(t) \leq \frac{b_p}{b_p - b_i} t \right] dt \qquad \text{(C.21)}$$

where $\lambda_i \, dt$ is the expected number of type i units that arrived during the time interval $(-t - dt, -t)$, and where $P_r[t \leq w_i(t) \leq t b_p/(b_p - b_i)]$ is the probability that a unit which arrived in that interval spends at least t and at most $t b_p/(b_p - b_i)$ seconds in the queue. Equation (C.21) can be written as

$$E(n_i) f_{ip} = \lambda_i \int_0^\infty (1 - P_r[w_i \leq t]) \, dt$$

$$- \lambda_i \int_0^\infty \left(1 - P_r \left[w_i \leq \frac{b_p}{b_p - b_i} t \right] \right) dt$$

$$= \lambda_i \int_0^\infty (1 - P_r[w_i \leq t]) \, dt$$

$$- \lambda_i \left(1 - \frac{b_i}{b_p} \right) \int_0^\infty (1 - P_r[w_i \leq \sigma]) \, d\sigma$$

where we have set

$$\sigma = \frac{b_p}{b_p - b_i} t$$

Now, as is well known[1] (for w_i a nonnegative random variable),

$$E(w_i) = \int_0^\infty (1 - P_r[w_i \leq x]) \, dx$$

Since, in our notation, $W_i = E(w_i)$, we obtain

$$E(n_i) f_{ip} = \lambda_i W_i - \lambda_i \left(1 - \frac{b_i}{b_p} \right) W_i$$

or

$$f_{ip} = \frac{\lambda_i W_i}{E(n_i)} \frac{b_i}{b_p}$$

[1] See, for example, Morse [20, p. 9].

But we know that

$$E(n_i) = \lambda_i W_i$$

Therefore,

$$f_{ip} = \frac{b_i}{b_p} \qquad \text{for all } i \leq p$$

Having derived expressions for f_{ip} and g_{ip}, we may now substitute for these quantities in Eq. (C.20) and obtain

$$W_p = \frac{W_0 + \sum_{i=p}^{P} \rho_i W_i + \sum_{i=1}^{p-1} \rho_i W_i b_i / b_p}{1 - \sum_{i=p+1}^{P} \rho_i (1 - b_p/b_i)}$$

If we now make use of Theorem 5.4, we can rewrite the above equation as

$$W_p = \frac{W_0/(1 - \rho) - \sum_{i=1}^{p-1} \rho_i W_i (1 - b_i/b_p)}{1 - \sum_{i=p+1}^{P} \rho_i (1 - b_p/b_i)}$$

which establishes Eq. (5.12) of Theorem 5.2.

Let us now show that Eq. (5.13) is indeed the solution to the set of recursively defined W_p, as expressed in Eq. (5.12). We proceed to do this by an inductive proof on p.

First, for $p = 1$, we get, from Eq. (5.13),

$$W_1 = \frac{W_0}{1 - \rho} \frac{1}{D_1} = \frac{W_0}{1 - \rho} \frac{1}{1 - \sum_{i=2}^{P} \rho_i (1 - b_1/b_i)}$$

which checks with the value of W_1 obtained from Eq. (5.12).

For $p = 2$, we get, from Eq. (5.13),

$$W_2 = \frac{W_0}{1 - \rho} \frac{1}{D_2} [1 + F_1(2)]$$

$$= \frac{W_0}{1 - \rho} \frac{1 - [\rho_1(1 - b_1/b_2)]/[1 - \sum_{i=2}^{P} \rho_i(1 - b_1/b_i)]}{1 - \sum_{i=3}^{P} \rho_i (1 - b_2/b_i)}$$

which checks with the value of W_2 obtained from Eq. (5.12).

Now, as is usual in an inductive proof, we assume that the solution holds for all $p \leq k$, and we show that this implies that the solution is correct for $p = k + 1$.   Let us therefore write the expression for W_{k+1} from Eq. (5.12), using the fact that W_k, W_{k-1}, . . . , W_1 may be evaluated from Eq. (5.13):

$$
\begin{aligned}
W_{k+1} &= \frac{W_0}{1 - \rho} \frac{1}{D_{k+1}} \Bigg\{ 1 - \sum_{i=1}^{k} \rho_i \left(1 - \frac{b_i}{b_{k+1}} \right) \frac{1}{D_i} \\
&\qquad\qquad \left[1 + \sum_{j=1}^{i-1} \sum_{0<i_1<\,\cdots\,<i_j<i} F_{i_1}(i_2) \,\cdots\, F_{i_j}(i) \right] \Bigg\} \\
&= \frac{W_0}{1 - \rho} \frac{1}{D_{k+1}} \Bigg\{ 1 + \sum_{i=1}^{k} F_i(k+1) \\
&\qquad\qquad \left[1 + \sum_{j=1}^{i-1} \sum_{0<i_1<\,\cdots\,<i_j<i} F_{i_1}(i_2) \,\cdots\, F_{i_j}(i) \right] \Bigg\}
\end{aligned}
$$

where we have taken the liberty of using the notation of Eqs. (5.14) and (5.15).   Now, comparing this last equation with the expression obtained for W_{k+1} from Eq. (5.13), we see that the induction proves the result if the following identity exists:

$$
\begin{aligned}
\sum_{i=1}^{k} F_i(k+1) &\left[1 + \sum_{j=1}^{i-1} \sum_{0<i_1<\,\cdots\,<i_j<i} F_{i_1}(i_2) \,\cdots\, F_{i_j}(i) \right] \\
&= \sum_{j=1}^{k} \sum_{0<i_1<\,\cdots\,<i_j<k+1} F_{i_1}(i_2) \,\cdots\, F_{i_j}(k+1)
\end{aligned}
$$

It is clear that both sides of this equation involve n-tuples of the F factors.   Therefore, to prove the validity of this expression, let us show that the same sets of n-tuples appear on both sides of the equation. First, for $n = 1$, we require that

$$
\sum_{i=1}^{k} F_i(k+1) = \sum_{i_1=1}^{k} F_{i_1}(k+1)
$$

which is obviously true.   Now, for $n > 1$, writing only the n-tuples for each side of the equation, we obtain

$$
\begin{aligned}
\sum_{i=1}^{k} F_i(k+1) &\sum_{0<i_1<\,\cdots\,<i_{n-1}<i} F_{i_1}(i_2) \,\cdots\, F_{i_{n-1}}(i) \\
&= \sum_{0<i_1<\,\cdots\,<i_n<k+1} F_{i_1}(i_2) \,\cdots\, F_{i_n}(k+1)
\end{aligned}
$$

If, on the right-hand side of this last equation, we separate out the summation involving i_n as follows:

$$\sum_{0<i_1<\,\cdots\,<i_n<k+1} F_{i_1}(i_2) \,\cdots\, F_{i_n}(k+1)$$

$$= \sum_{i_n=1}^{k} F_{i_n}(k+1) \sum_{0<i_1<\,\cdots\,<i_{n-1}<i_n} F_{i_1}(i_2) \,\cdots\, F_{i_{n-1}}(i_n)$$

we find that the n-tuples do indeed agree (i.e., let $i_n = i$ in this last expression). Thus we have proven the validity of Eq. (5.13), and this completes the proof of Theorem 5.2.

C.5   Theorem 5.3 and Its Proof

Theorem 5.3

For the delay-dependent priority system with preemption and $0 \le \rho < 1$,

$$W_p = \frac{1}{1 - \displaystyle\sum_{i=p+1}^{P} \rho_i(1 - b_p/b_i)} \left[\frac{W_0}{1-\rho} + \sum_{i=p+1}^{P} \frac{\rho_i}{\mu_p}\left(1 - \frac{b_p}{b_i}\right) \right.$$

$$\left. - \sum_{i=1}^{p-1} \frac{\rho_i}{\mu_i}\left(1 - \frac{b_i}{b_p}\right) - \sum_{i=1}^{p-1} \rho_i W_i \left(1 - \frac{b_i}{b_p}\right) \right]$$

PROOF: Here we use notation very similar to that used in the proof of Theorem 5.2, except that all quantities will refer to time spent in the queue plus time spent in the service facility, instead of just time spent in the queue, as was the case for Theorem 5.2.

Following through with almost identical arguments, we arrive at the following expressions:

$$E(n_i) = \lambda_i T_i$$
$$E(m_i) = \lambda_i T_p$$
$$f_{ip} = \begin{cases} \dfrac{b_i}{b_p} & i \le p \\ 1 & i \ge p \end{cases}$$
$$g_{ip} = \begin{cases} 0 & i \le p \\ 1 - \dfrac{b_p}{b_i} & i \ge p \end{cases}$$

where n_i is now defined as the total number of type i units which were present in the system (queue plus service facility) when the tagged unit arrived, and m_i is defined as the total number of type i units which enter the system while the tagged unit is in the system.

The expression for T_p is therefore

$$T_p = \frac{1}{\mu_p} + \sum_{i=1}^{P} \frac{\lambda_i T_i}{\mu_i} f_{ip} + \sum_{i=1}^{P} \frac{\lambda_i T_p}{\mu_i} g_{ip}$$

This equation is obtained from reasoning quite similar to that used in forming Eq. (C.19). Now, using the expressions for f_{ip} and g_{ip} and remembering that $W_i + 1/\mu_i = T_i$, we obtain

$$W_p = T_p - \frac{1}{\mu_p} = \sum_{i=1}^{p} \rho_i \left(W_i + \frac{1}{\mu_i} \right) \frac{b_i}{b_p} + \frac{W_0}{1 - \rho}$$

$$- \sum_{i=1}^{p} \rho_i \left(W_i + \frac{1}{\mu_i} \right) + \sum_{i=p+1}^{P} \rho_i \left(W_p + \frac{1}{\mu_p} \right) \left(1 - \frac{b_p}{b_i} \right)$$

We have applied Theorem 5.4 to obtain this last expression. Solving for W_p and collecting terms, we obtain, finally,

$$W_p = \frac{1}{1 - \sum\limits_{i=p+1}^{P} \rho_i(1 - b_p/b_i)} \left[\frac{W_0}{1 - \rho} + \sum_{i=p+1}^{P} \frac{\rho_i}{\mu_p} \left(1 - \frac{b_p}{b_i} \right) \right.$$

$$\left. - \sum_{i=1}^{p-1} \frac{\rho_i}{\mu_i} \left(1 - \frac{b_i}{b_p} \right) - \sum_{i=1}^{p-1} \rho_i W_i \left(1 - \frac{b_i}{b_p} \right) \right]$$

which is the same as Eq. (5.16) and so proves Theorem 5.3.

Note that $E(n)$, the expected number of units in the system, is

$$E(n) = \sum_{p=1}^{P} E(n_p) = \sum_{p=1}^{P} \lambda_p T_p$$

C.6 Theorem 5.5 and Its Proof

Theorem 5.5

The expected value of the total time T_n spent in the late-arrival system by a message whose service time is nQ seconds is

$$T_n = \frac{nQ}{1 - \rho} - \frac{\lambda Q^2}{1 - \rho} \left[1 + \frac{(1 - \sigma\alpha)(1 - \alpha^{n-1})}{(1 - \sigma)^2(1 - \rho)} \right]$$

where $\alpha = \sigma + \lambda Q$

PROOF.   Let us first prove Eq. (5.24), which gives the expected value of the distribution r_k, where

$$r_k = (1 - a)a^k$$

Clearly,

$$E = \sum_{k=0}^{\infty} kr_k$$

$$E = \frac{a}{1 - a}$$

But

$$a = \frac{\rho\sigma}{1 - \lambda Q}$$

and so

$$E = \frac{\rho\sigma}{1 - \rho}$$

where we have used the definition of $\rho = \lambda Q/(1 - \sigma)$ as before.   This establishes Eq. (5.24).

Now for the theorem.   The arguments which follow are based exclusively on expected values.   Consider the arrival of a unit (the tagged unit) whose service time is nQ sec.   Let D_i be the expected value of the delay (or time spent) between the completion of its $(i - 1)$st service interval and the completion of its ith service interval.   (We assume that the completion of its zeroth service interval occurs at its time of arrival.)   Clearly then, T_n, the expected value of the total time spent in the system by such a unit, will be

$$T_n = \sum_{i=1}^{n} D_i$$

Let us further define N_i as the expected number of service intervals (or units which are serviced) between the completion of the $(i - 1)$st and ith service intervals of the tagged unit; i.e.,

$$N_i = \frac{D_i}{Q}$$

and so

$$T_n = Q \sum_{i=1}^{n} N_i \tag{C.22}$$

We now derive a general form for N_i.   Upon its arrival to the system, the tagged unit finds a certain number of units in the queue, the expected value of which is E by definition.   Note that the service facility is empty whenever a new unit enters the system.   Thus,

$$N_1 = E + 1$$

Unity is added to account for the service interval used up in serving the tagged unit's first time interval.   Now, each of these E units will remain in the system with probability σ, and so $\sigma(N_1 - 1)$ of them will contribute to N_2. In addition, during the time $Q(N_1 - 1)$ devoted to servicing these E units, we expect λ new units to arrive per second, and so we must add $\lambda Q(N_1 - 1)$ units to N_2. Besides all this, for $n > 1$, we must add one more unit (the tagged unit itself) to N_2, giving

$$N_2 = \sigma(N_1 - 1) + \lambda Q(N_1 - 1) + 1$$
$$= (\sigma + \lambda Q) E + 1$$

In calculating N_3, we see that a fraction σ of the units which were serviced before the second time interval of the tagged unit will remain in the system, that is, $\sigma(N_2 - 1)$. In addition, during the time $Q(N_2 - 1)$ devoted to servicing these units, $\lambda Q(N_2 - 1)$ new units will arrive. Also, for $n > 2$, we must add one more unit (the tagged unit again) to N_3. However, we now notice a new effect entering, namely, the presence of a unit which arrived (with probability λQ) at the conclusion of the first service interval of the tagged unit. This additional unit was placed in back of the tagged unit when it arrived and therefore did not appear in N_2. However, from now on it will appear as an additional λQ added to each N_i for $i \geq 3$. Thus,

$$N_3 = \sigma(N_2 - 1) + \lambda Q(N_2 - 1) + 1 + \lambda Q$$
$$= (\sigma + \lambda Q)^2 E + \lambda Q + 1$$

For N_i, we merely repeat the arguments used in finding N_3, with the substitutions N_i for N_3 and N_{i-1} for N_2. This gives us, for $i = 3$, $4, \ldots, n$,

$$N_i = \sigma(N_{i-1} - 1) + \lambda Q(N_{i-1} - 1) + \lambda Q + 1$$
$$= (\sigma + \lambda Q)(N_{i-1} - 1) + \lambda Q + 1 \tag{C.23}$$

Now, letting $\alpha = \sigma + \lambda Q$, we assert that

$$N_i = \alpha^{i-1} E + \lambda Q \sum_{j=0}^{i-3} \alpha^j + 1 \tag{C.24}$$

is the solution of Eq. (C.23) for $i = 3, 4, \ldots, n$. Let us prove this by induction. Clearly, it holds for $i = 3$. Assuming its validity for N_{i-1}, we show its validity for N_i as follows:

$$N_i = \alpha(N_{i-1} - 1) + \lambda Q + 1$$
$$= \alpha \left(\alpha^{i-2} E + \lambda Q \sum_{j=0}^{i-4} \alpha^j \right) + \lambda Q + 1$$
$$= \alpha^{i-1} E + \lambda Q \sum_{j=0}^{i-4} \alpha^{j+1} + \lambda Q + 1$$
$$= \alpha^{i-1} E + \lambda Q \sum_{j=0}^{i-3} \alpha^j + 1$$

This proves the assertion.   Now if we take the usual definition of

$$\sum_{i=a}^{b} x_i = 0 \qquad \text{for } b < a$$

we see that Eq. (C.24) also holds for $i = 1, 2$.   Thus we find that N_i ($i = 1, 2, \ldots, n$), as given by Eq. (C.24), is the general form we were seeking.   Recognizing that, for $0 \leq b$ and $|x| < 1$,

$$\sum_{k=a}^{b+a} x^k = x^a \frac{1 - x^{b+1}}{1 - x} \tag{C.25}$$

we can readily evaluate N_i.   First let us show that $x < 1$ (that is, that $\alpha < 1$).   This is easily done by recalling that we are dealing with systems in equilibrium (steady state).   This implies that $\rho < 1$.   Substituting for ρ, we get

$$\rho = \frac{\lambda Q}{1 - \sigma} < 1$$

or

$$\lambda Q + \sigma < 1$$

which shows, of course, that $\alpha < 1$.
   We now use Eq. (C.25) in Eq. (C.24) to obtain

$$N_i = \begin{cases} E + 1 & i = 1 \\ \alpha^{i-1} E + \lambda Q \dfrac{1 - \alpha^{i-2}}{1 - \alpha} + 1 & i = 2, 3, \ldots, n \end{cases} \tag{C.26}$$

Substituting for E and collecting terms, we find

$$N_i = \begin{cases} \dfrac{1 - \lambda Q}{1 - \rho} & i = 1 \\ \dfrac{1}{1 - \rho} - \dfrac{\rho(1 - \sigma\alpha)}{1 - \rho} \alpha^{i-2} & i = 2, 3, \ldots, n \end{cases} \tag{C.27}$$

We are now in a position to evaluate T_n in Eq. (C.22) by substituting Eq. (C.27).   Performance of the required operations and recognition that $1 - \alpha = (1 - \sigma)(1 - \rho)$ lead us to

$$T_n = \frac{nQ}{1 - \rho} - \frac{\lambda Q^2}{1 - \rho} \left[1 + \frac{(1 - \sigma\alpha)(1 - \alpha^{n-1})}{(1 - \sigma)^2(1 - \rho)} \right]$$

which completes the proof of Theorem 5.5.

C.7 Theorem 5.6 and Its Proof

Theorem 5.6

The expected value T_n of the total time spent in the early-arrival system by a message whose service time is nQ seconds is

$$T_n = \frac{nQ}{1 - \rho} - \rho Q - \frac{\lambda Q^2 \rho}{1 - \rho}\left[1 + \frac{(1 - \sigma\alpha)(1 - \alpha^{n-1})}{(1 - \sigma)^2(1 - \rho)}\right]$$

PROOF. Let us first establish the distribution of r_k as given by Eq. (5.27). Using methods similar to those of Morse [20], we derive the following equilibrium relationships among the r_k:

$$\lambda Q r_0 = (1 - \sigma)(1 - \lambda Q)r_1$$
$$[(1 - \sigma)(1 - \lambda Q) + \lambda Q\sigma]r_1 = \lambda Q r_0 + (1 - \sigma)(1 - \lambda Q)r_2$$
$$[(1 - \sigma)(1 - \lambda Q) + \lambda Q\sigma]r_k = \lambda Q\sigma r_{k-1} + (1 - \sigma)(1 - \lambda Q)r_{k+1} \qquad k \geq 2$$

As before, let

$$a = \frac{\rho\sigma}{1 - \lambda Q}$$

$$\rho = \frac{\lambda Q}{1 - \sigma}$$

It is then a simple matter to show that the solution to the above equations is

$$r_k = \begin{cases} 1 - \rho & k = 0 \\ \dfrac{1 - \rho}{\sigma}\, a^k & k = 1, 2, \ldots \end{cases}$$

which proves Eq. (5.27).

The expected value of the distribution of r_k is

$$E' = \sum_{k=0}^{\infty} k r_k = \frac{1 - \rho}{\sigma}\, a \sum_{k=0}^{\infty} k a^{k-1} = \frac{(1 - \rho)a}{(1 - a)^2\sigma}$$

But $1 - a = 1 - \dfrac{\rho\sigma}{1 - \lambda Q} = \dfrac{1 - \lambda Q - \sigma + \sigma(1 - \rho)}{1 - \lambda Q} = \dfrac{1 - \rho}{1 - \lambda Q}$

Thus, $$E' = \frac{\rho}{1 - \rho}\,(1 - \lambda Q)$$

which proves Eq. (5.28).

Now for the theorem. The arguments needed here are quite similar to those used in Theorem 5.5 and will therefore be considerably shortened. In particular, we define T_n, D_i, and N_i as previously, thereby establishing Eq. (C.22) again. Let us now derive a general form for N_i. Upon entering the system, the tagged unit finds E' units in the system. Now, if there is a unit in the service facility (which occurs with probability $1 - r_0 = \rho$), only E' minus the expected value of the number in the service facility will contribute to N_1 (since any unit in service must be on the verge of being ejected from service). This expected value is just

$$0(r_0) + 1(1 - r_0) = \rho$$

and so
$$N_1 = E' - \rho + 1$$

where the $+1$ term is due to the tagged unit itself. Following the same reasoning as in Theorem 5.5, we find that

$$N_2 = \lambda Q(N_1) + \sigma(N_1 - 1) + \sigma\rho + 1$$

where the $\sigma\rho$ term is due to the unit (if there is one) found in service at the time of arrival of the tagged unit. Using the same type of argument, we find, for $i > 2$,

$$N_i = \lambda Q(N_{i-1}) + \sigma(N_{i-1} - 1) + 1$$

where we omit the $\sigma\rho$ term, since it is fully accounted for in N_2. We assert that the solution to this set of equations is

$$N_i = \begin{cases} E' + 1 - \rho & i = 1 \\ \alpha^{i-1}E' + \alpha^{i-2}\lambda Q(1 - \rho) + \lambda Q \sum_{j=0}^{i-3} \alpha^j + 1 & i > 1 \end{cases} \quad \text{(C.28)}$$

where
$$\alpha = \sigma + \lambda Q$$

That this is indeed the solution is easily shown by induction on i. Clearly, it is true for $i = 1, 2$. Now, assuming its validity for N_{i-1}, we show its validity for N_i as follows:

$$\begin{aligned} N_i &= \alpha N_{i-1} + 1 - \sigma \\ &= \alpha \left[\alpha^{i-2}E' + \alpha^{i-3}\lambda Q(1 - \rho) + \lambda Q \sum_{j=0}^{i-4} \alpha^j + 1 \right] + 1 - \sigma \\ &= \alpha^{i-1}E' + \alpha^{i-2}\lambda Q(1 - \rho) + \lambda Q \sum_{j=1}^{i-3} \alpha^j + \alpha + 1 - \sigma \\ &= \alpha^{i-1}E' + \alpha^{i-2}\lambda Q(1 - \rho) + \lambda Q \sum_{j=0}^{i-3} \alpha^j + 1 \end{aligned}$$

which proves the assertion. Substitution for E', performance of the indicated summation, and collection of terms give us

$$N_i = \begin{cases} \dfrac{\rho}{1-\rho}(1-\lambda Q) + 1 - \rho & i = 1 \\[2mm] \dfrac{1}{1-\rho} - \dfrac{\rho^2(1-\sigma\alpha)}{1-\rho}\alpha^{i-2} & i = 2, 3, \ldots, n \end{cases} \qquad \text{(C.29)}$$

We are now in a position to evaluate T_n in Eq. (C.22) by substituting Eq. (C.29). Performing the required operations leads us to

$$T_n = \frac{nQ}{1-\rho} - \rho Q - \frac{\lambda Q^2 \rho}{1-\rho}\left[1 + \frac{(1-\sigma\alpha)(1-\alpha^{n-1})}{(1-\sigma)^2(1-\rho)}\right]$$

which proves Theorem 5.6.

C.8 Theorem 5.7 and Its Proof

Theorem 5.7

The expected value T_n of the total time spent in the strict first-come first-served system by a message whose service time is nQ seconds is

$$T_n = \frac{QE}{1-\sigma} + nQ$$

where

$$E = \frac{\rho\sigma}{1-\rho}$$

PROOF: Let us first consider the late-arrival system. Arguing on an expected-value basis, we recognize that, upon entry, the tagged unit finds $E = \rho\sigma/(1-\rho)$ units in the system. Each unit in the queue has an expected service time of $Q/(1-\sigma)$. Now, as far as the unit in service is concerned, we appeal to the discussion leading up to Eq. (C.2) for the exponential distribution and assert that the same type of result holds. That is, we assert that the expected additional service time for the unit in service is $Q/(1-\sigma)$ (given that more service is required). It may be seen that this assertion is true by recognizing that the geometric distribution is the discrete counterpart of the exponential distribution. Thus, each of the E units in the system (queue plus service) will delay the tagged unit by $Q/(1-\sigma)$ seconds, and this unit will spend nQ seconds in service itself. Hence, for the late-arrival system,

$$T_n = Q\frac{\rho\sigma}{(1-\sigma)(1-\rho)} + nQ$$

which proves Theorem 5.7 for the late-arrival system.

For the early-arrival system, we recognize that, upon entry, the tagged unit finds $E' = [\rho/(1 - \rho)](1 - \lambda Q)$ units in the system. Now as before, given that the unit in service requires additional service, the expected value of this additional service is $Q/(1 - \sigma)$ seconds. But now we cannot be sure (as we were in the late-arrival system) that the unit in service will require more service; i.e., with probability σ the unit in service will remain for more service. Also, the expected number of units in service is merely ρ (that is, the probability of finding one unit in service), and so the delay suffered by the tagged unit due to the unit in service is $\rho\sigma Q/(1 - \sigma)$. Each of the units in the queue (the expected number of which is $E' - \rho$) will, on the average, delay the tagged unit by $Q/(1 - \sigma)$ seconds. In addition, the tagged unit will itself spend nQ seconds in service. Hence, for the early-arrival system,

$$T_n = (E' - \rho) \frac{Q}{1 - \sigma} + \rho\sigma \frac{Q}{1 - \sigma} + nQ$$

$$= \frac{Q}{1 - \sigma} [E' - \rho(1 - \sigma)] + nQ$$

Note that

$$\Delta = E' - E = \rho(1 - \sigma) = [\rho/(1 - \rho)](1 - \lambda Q) - \rho\sigma/(1 - \rho)$$

thus
$$T_n = Q \frac{\rho\sigma}{(1 - \sigma)(1 - \rho)} + nQ$$

which proves Theorem 5.7 for the early-arrival system.

appendix D

Theorems and Proofs
for Chapter 6

D.1 Definitions and Derived Expressions for $\bar{n}$

We define x_n as the node visited by a message during its nth step in the Markov process. Thus, x_n determines the system state at step n. Further, let node N represent the destination or absorbing node for our message. We define:

r_n = P_r[message reaches node N on exactly nth step]
s_n = P_r[message reaches node N on or before nth step]
g_n = P_r[message reaches node N on exactly nth step, given that it did not reach node N before nth step]

Also define, for completeness, $s_n = 0$ for $n < 0$. As usual, $\bar{n}$ is the expected number of steps to reach state N:

$$\bar{n} = \sum_{n=0}^{\infty} n r_n \tag{D.1}$$

We now proceed to derive two other forms for $\bar{n}$. We consider only finite Markov processes which consist of exactly one closed set (namely, state N itself).[1] Since the message must eventually be absorbed in

[1] See Feller [18, Sec. XV. 4].

node N, we see that

$$\sum_{n=0}^{\infty} r_n = 1 \tag{D.2}$$

Thus states $0, 1, 2, \ldots, N-1$ are transient states. Furthermore, by definition,

$$s_n = \sum_{m=0}^{n} r_m \tag{D.3}$$

Now, Eq. (D.1) may be written as

$$\bar{n} = \sum_{n=1}^{\infty} r_n \sum_{m=0}^{n-1} 1$$

Inverting the order of summation, we obtain

$$\bar{n} = \sum_{m=0}^{\infty} \sum_{n=m+1}^{\infty} r_n$$
$$= \sum_{m=0}^{\infty} \left(\sum_{n=0}^{\infty} r_n - \sum_{n=0}^{m} r_n \right)$$

Using Eqs. (D.2) and (D.3), we get, as an alternative form[1] for $\bar{n}$,

$$\bar{n} = \sum_{n=0}^{\infty} (1 - s_n) \tag{D.4}$$

From its definition, we see that

$$g_n = \frac{r_n}{1 - s_{n-1}}$$

and since, from Eq. (D.3),

$$r_n = s_n - s_{n-1}$$

we get

$$g_n = \frac{s_n - s_{n-1}}{1 - s_{n-1}}$$

or

$$1 - g_n = \frac{1 - s_n}{1 - s_{n-1}}$$

It is now clear that

$$\prod_{m=0}^{n} (1 - g_m) = \prod_{m=0}^{n} \frac{1 - s_m}{1 - s_{m-1}} = 1 - s_n \tag{D.5}$$

Thus, from Eqs. (D.4) and (D.5), we obtain

$$\bar{n} = \sum_{n=0}^{\infty} (1 - g_0)(1 - g_1) \cdots (1 - g_n) \tag{D.6}$$

---

[1] Feller derives this same expression in a different manner [18, p. 249].

Thus, we see that knowledge of either the set r_n, the set s_n, or the set g_n is sufficient to obtain $\bar{n}$ as given by Eq. (D.1), (D.4), or (D.6), respectively.

D.2   Theorem 6.1 and Its Proof

Theorem 6.1

The average path length $\bar{n}_i$ from node i to node N for any finite-dimensional irreducible Markov process whose probability transition matrix is a circulant matrix [see Eq. (6.2)] is

$$\bar{n}_i = \sum_{r=1}^{N} \frac{1 - \theta^{r(i+1)}}{1 - \sum_{s=0}^{N} q_s \theta^{sr}}$$

where $i = 0, 1, 2, \ldots, N$, and where θ is the $(N+1)$th primitive root of unity; i.e.,

$$\theta = e^{2\pi j/(N+1)}$$

and $$j = \sqrt{-1}$$

PROOF.   Let[1]

$$p_{ij}(n) = \Pr[x_n = j | x_0 = i]$$

and $f_{ij}(n) = \Pr[x_n = j, x_m \neq j \ (m = 1, 2, \ldots, n-1) | x_0 = i]$

That is, $p_{ij}(n)$ is the probability that the message is in node j exactly n steps after being in node i; $f_{ij}(n)$ is the probability that this event occurs at step n for the first time (i.e., it is the first passage probability).   We also define the two *generating functions* for those probabilities as

$$P_{ij}(t) = \sum_{n=0}^{\infty} p_{ij}(n)t^n \qquad -1 < t < 1 \tag{D.7}$$

$$F_{ij}(t) = \sum_{n=0}^{\infty} f_{ij}(n)t^n \qquad -1 < t < 1 \tag{D.8}$$

Now, it is clear that if the message is in node j after n steps, it must have reached this node for the first time on the n_1th step ($n_1 \leq n$) and must have gone from node j back to node j (perhaps many times) in $n - n_1$ steps.   This is expressed in the following equation:

$$p_{ij}(n) = \sum_{n_1=0}^{n} f_{ij}(n_1)p_{jj}(n - n_1)$$

[1] See Sec. 6.1 for the definition of x_n.

Now, multiplying by t^n and summing over all n, we get

$$\sum_{n=0}^{\infty} p_{ij}(n)t^n = \sum_{n=0}^{\infty} \sum_{n_1=0}^{n} f_{ij}(n_1)p_{jj}(n - n_1)t^{n_1}t^{n-n_1}$$

Interchanging orders of summation and setting $n_2 = n - n_1$, we obtain

$$\sum_{n=0}^{\infty} p_{ij}(n)t^n = \sum_{n_1=0}^{\infty} \sum_{n_2=0}^{\infty} f_{ij}(n_1)p_{jj}(n_2)t^{n_1}t^{n_2} \qquad (D.9)$$

We recognize that Eq. (D.9) is merely[1]

$$P_{ij}(t) = F_{ij}(t)P_{jj}(t)$$

or

$$F_{ij}(t) = \frac{P_{ij}(t)}{P_{jj}(t)} \qquad (D.10)$$

We now define the circulant matrix P (finite-dimensional and irreducible) as given by Eq. (6.2). For this matrix, Feller [18] has shown that

$$p_{ij}(n) = \frac{1}{N+1} \sum_{r=0}^{N} \theta^{r(i-j)} \Big(\sum_{s=0}^{N} q_s\theta^{sr} \Big)^n \qquad (D.11)$$

where θ is the $(N + 1)$th primitive root of unity as expressed in Eq. (6.4). We now form $P_{ij}(t)$ for this matrix by applying the transformation of Eq. (D.7) to Eq. (D.11) to obtain

$$P_{ij}(t) = \frac{1}{N+1} \sum_{n=0}^{\infty} \sum_{r=0}^{N} \theta^{r(i-j)} \Big(t \sum_{s=0}^{N} q_s\theta^{sr} \Big)^n \qquad (D.12)$$

or

$$P_{ij}(t) = \frac{1}{N+1} \sum_{r=0}^{N} \frac{\theta^{r(i-j)}}{1 - t \sum_{s=0}^{N} q_s\theta^{sr}} \qquad (D.13)$$

We are able to sum the geometric series of Eq. (D.12), since

$$\Big| t \sum_{s=0}^{N} q_s\theta^{sr} \Big| \leq \Big| t \sum_{s=0}^{N} q_s \Big| = |t| < 1$$

Thus, substituting Eq. (D.13) into Eq. (D.10), we obtain

$$F_{ij}(t) = \frac{\displaystyle\sum_{r=0}^{N} \theta^{r(i-j)}/(1 - tS_r)}{\displaystyle\sum_{r=0}^{N} 1/(1 - tS_r)}$$

[1] See Kemperman [54].

or

$$F_{ij}(t) = \frac{1 + (1 - t) \sum\limits_{r=1}^{N} \theta^{r(i-j)}/(1 - tS_r)}{1 + (1 - t) \sum\limits_{r=1}^{N} 1/(1 - tS_r)}$$

where

$$S_r = \sum_{s=0}^{N} q_s \theta^{sr}$$

Upon setting $j = N$, we establish Eq. (6.5).

We now define

$\bar{n}_i$ = expected number of steps to enter destination node N for first time, given that message originated in node i

That is,

$$\bar{n}_i = \sum_{n=0}^{\infty} n f_{iN}(n)$$

and since we have a finite irreducible chain, we are assured of the existence of $\bar{n}_i$. From Eq. (D.8), we see that

$$\frac{\partial F_{iN}(t)}{\partial t} = \sum_{n=0}^{\infty} n f_{iN}(n) t^{n-1}$$

We know that this series converges for $-1 < t < 1$. At $t = 1$, the series becomes $\bar{n}_i$ by definition. Since $\bar{n}_i$ exists, we recognize that $\partial F_{iN}(t)/\partial t$ must be continuous in the closed interval $-1 \le t \le 1$ (see Feller [18], p. 249, for a similar extension of the region of convergence). Thus, after differentiating Eq. (6.5), setting $t = 1$, and noting that $\theta^{i-N} = \theta^{i+1}$, we obtain

$$\bar{n}_i = \sum_{r=1}^{N} \frac{1 - \theta^{r(i+1)}}{1 - S_r}$$

which completes the proof of Theorem 6.1.

D.3   Proof of Eqs. (6.6) and (6.7)

Making use of Eq. (6.3), we form

$$\bar{n} = \frac{1}{N} \sum_{i=0}^{N} \bar{n}_i = \frac{1}{N} \sum_{i=0}^{N} \sum_{r=1}^{N} \frac{1 - \theta^{r(i+1)}}{1 - S_r}$$

$$= \frac{1}{N} \sum_{r=1}^{N} \frac{N + 1 - \sum\limits_{i=0}^{N} \theta^{r(i+1)}}{1 - S_r}$$

But, for $N \geq 1$, we note that

$$\sum_{i=0}^{N} \theta^{r(i+1)} = \begin{cases} N+1 & \text{for } r = 0, N+1 \\ 0 & \text{for } r \neq 0, N+1 \end{cases}$$

and since we have $r = 1, 2, \ldots, N$, we obtain

$$\bar{n} = \frac{N+1}{N} \sum_{r=1}^{N} \frac{1}{1 - S_r}$$

which proves Eq. (6.6).

Furthermore, we form

$$\bar{n}' = \sum_{i=0}^{N-1} q_{i+1} \bar{n}_i + q_0 \bar{n}_N$$

$$= \sum_{i=0}^{N-1} q_{i+1} \sum_{r=1}^{N} \frac{1 - \theta^{r(i+1)}}{1 - S_r} + q_0 \cdot 0$$

$$= \sum_{r=1}^{N} \frac{1 - q_0 - \sum_{i=0}^{N-1} q_{i+1} \theta^{r(i+1)}}{1 - S_r}$$

or

$$\bar{n}' = \sum_{r=1}^{N} \frac{1 - S_r}{1 - S_r} = N$$

This establishes Eq. (6.7).

D.4   Theorem D.1 and Its Proof

Theorem D.1

Given a two-channel service facility of total capacity C, Poisson arrivals with mean rate λ, message lengths distributed exponentially with mean length $1/\mu$, and the restriction that no channel be idle if a message is waiting in the queue, then, for an arbitrarily chosen number $0 \leq \pi_1 \leq 1$, it is *not* possible to find a queue discipline and an assignment of the two channel capacities (the sum being C) such that

P_r[entering message is transmitted on first channel] $= \pi_1$ (D.14)

for all $0 \leq \rho < 1$, where $\rho = \lambda/\mu C$.

PROOF: We prove this theorem by considering two limiting cases. Suppose $\rho \to 0$. Then P_0 (the probability that in the steady state the system is empty) approaches 1. In such a case, an entering message (which will, with probability arbitrarily close to 1, find an empty system) must be assigned to channel 1 with probability π_1 (and to channel 2 with probability $\pi_2 = 1 - \pi_1$) if Eq. (D.14) is to hold.

Now suppose $\rho \to 1$; then P_0 and P_1 (the probability of one message in the system) both approach 0. Therefore, the channel capacity C_1 assigned to channel 1 (which implies $C - C_1 = C_2$ for channel 2) must be chosen so that

$$\alpha \equiv P_r[\text{channel 1 empties before channel 2}|\text{both channels busy}] = \pi_1$$

That is, with probability arbitrarily close to 1, a message entering the node will be forced to join a queue and, when it reaches the head of the queue, will find both channels busy. If this message is to be transmitted over channel 1 with probability π_1, the channel capacity assignments must result in $\alpha = \pi_1$. Note that we have taken advantage of the fact that messages with exponentially distributed lengths exhibit no memory as regards their transmission time [see Eq. (C.2)]. Now,

$$\alpha = \int_{t=0}^{t=\infty} P_r[\text{channel 1 empties in } (t, t + dt)|\text{both busy at time zero}]$$
$$\times P_r[\text{channel 2 is not yet empty by } t|\text{both busy at time zero}]$$
$$= \int_0^\infty \mu C_1 e^{-\mu C_1 t - \mu C_2 t} \, dt$$
$$= \frac{\mu C_1}{\mu C_1 + \mu C_2} = \frac{C_1}{C}$$

But
$$\alpha = \pi_1$$
Therefore
$$C_1 = \pi_1 C$$
and
$$C_2 = \pi_2 C = (1 - \pi_1)C$$

These two limiting cases, $\rho \to 0$ and $\rho \to 1$, have constrained the construction of our system completely.

Now, for any $0 \le \rho < 1$, let

$r_1 = P_r[\text{incoming message is eventually transmitted on channel 1}]$
$P_n = P_r[\text{finding } n \text{ messages in system in steady state}]$

Then, clearly,

$$r_1 = \pi_1 P_0 + q_{21} P_1 + \sum_{n=2}^\infty \pi_1 P_n \tag{D.15}$$

where $q_{i1} = P_r[\text{channel } i \text{ is busy}|\text{only one channel is busy}]$
For q_{21}, we write the forward Chapman-Kolmogorov equations (see Sec. A.1):

$$q_{21}(t + dt) = \frac{P_0(t)}{P_1(t)} \lambda \pi_2 \, dt + \frac{P_2(t)}{P_1(t)} \mu C_1 \, dt + q_{21}(t)(1 - \lambda \, dt - \mu C_2 \, dt)$$

Assuming a steady-state distribution, we get

$$0 = \frac{P_0}{P_1} \lambda \pi_2 + \frac{P_2}{P_1} \mu C_1 - (\lambda + \mu C_2) q_{21} \tag{D.16}$$

Now, since this system satisfies the hypothesis of the birth and death process considered earlier, we apply Eq. (A.1), with $d_1 = \mu \bar{E}_1$, $d_n = \mu C$ ($n \geq 2$), and $b_n = \lambda$, to obtain

$$P_n = \begin{cases} \dfrac{C}{\bar{E}_1} \rho^n P_0 & n \geq 1 \\ P_0 & n = 0 \end{cases} \tag{D.17}$$

where

$$\rho = \frac{\lambda}{\mu C}$$

and[1]
$$\begin{aligned} \bar{E}_1 &= E(\text{capacity in use}|\text{one channel is busy}) \\ &= C_1 q_{11} + C_2 q_{21} \end{aligned}$$

Recall that $C_1 = \pi_1 C$ and $C_2 = \pi_2 C = (1 - \pi_1)C$. Thus, Eq. (D.16) becomes

$$q_{21} = \frac{\mu \bar{E}_1 \pi_2 + \lambda \pi_1}{\lambda + \mu C \pi_2} \tag{D.18}$$

Similarly,
$$q_{11} = \frac{\mu \bar{E}_1 \pi_1 + \lambda \pi_2}{\lambda + \mu C \pi_1}$$

Now, forming the equation

$$q_{11} + q_{21} = 1$$

we obtain, after some algebra,

$$\begin{aligned} \mu \bar{E}_1 &= \frac{\mu C (\mu C + 2\lambda)}{2\mu C + \lambda/\pi_1 \pi_2} \\ &= \frac{\mu C (1 + 2\rho)}{2 + \rho/\pi_1 \pi_2} \end{aligned}$$

We may now write Eq. (D.18) as

$$q_{21} = \frac{[\mu C (1 + 2\rho)/(2 + \rho/\pi_1 \pi_2)]\pi_2 + \lambda \pi_1}{\lambda + \mu C \pi_2}$$

Simplifying, we obtain

$$q_{21} = \frac{\pi_1(\pi_2 + \rho)}{2\pi_1 \pi_2 + \rho} \tag{D.19}$$

Returning to Eq. (D.15), we observe that r_1 can equal π_1 only if $q_{21} = \pi_1$. Equation (D.19) shows that this is *not* the case, demon-

[1] See Theorem 4.1.

strating that Eq. (D.14) cannot hold for an arbitrary π_1 and proving the theorem.

However, it can be seen from Eq. (D.19) that $q_{21} = \pi_1$ for $\pi_1 = 0$, $\frac{1}{2}$, and 1 only. Let us now form r_1 from Eqs. (D.15) and (D.17):

$$r_1 = P_0\left(\pi_1 + \frac{\lambda q_{21}}{\mu \bar{E}_1} + \frac{\pi_1 C}{\bar{E}_1} \sum_{n=2}^{\infty} \rho^n\right)$$

where P_0 is found from Eq. (D.17) by requiring

$$\sum_{n=0}^{\infty} P_n = 1$$

After substituting and simplifying, we get

$$r_1 = \pi_1 \frac{\pi_1 \rho^2 + (1 - \pi_1{}^2)\rho + \pi_1 \pi_2}{(1 - 2\pi_1 \pi_2)\rho^2 + 3\pi_1 \pi_2 \rho + \pi_1 \pi_2}$$

Figure D.1 shows a plot of r_1 as a function of ρ, with π_1 as a parameter. Note that the variation of r_1 is not too great. This illustrates the fact that, although Theorem D.1 is stated as a negative result, its proof demonstrates a positive result, namely, that the variation of r_1 is not excessive. The arrangement which leads to this behavior is one in which the channel capacity is divided between the two channels in

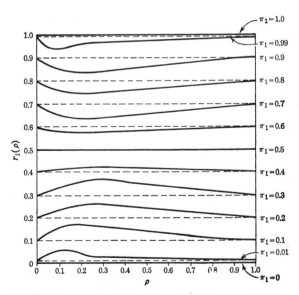

Fig. D.1.   *Variation of r_1 with ρ.*

proportion to the desired probabilities of using each channel (that is, $C_i = \pi_i C$); a message finding both channels empty chooses channel i with probability π_i.

Corollary

For the conditions of Theorem D.1, with K channels and $\pi_1 = \pi_2 = \cdots = \pi_K = 1/K$, it is possible to find a queueing discipline and a channel capacity assignment such that

P_r[entering message is transmitted over ith channel] $= \dfrac{1}{K}$ for all

$0 \le \rho < 1$.

PROOF: In proving Theorem D.1, it was shown that a system could be found to realize $\pi_1 = \frac{1}{2}$. This result also follows directly from the complete symmetry of the two channels. The proof of the present corollary follows trivially from recognition, once again, of the complete symmetry of the K channels.

D.5   Theorem 6.2 and Its Proof

Theorem 6.2

For the K-connected net,

$$P_n = \begin{cases} P_0(\bar{n}\rho)^n \dfrac{K^n}{n!} & n = 0, 1, 2, \ldots, K \\[2mm] P_0(\bar{n}\rho)^n \dfrac{K^K}{K!} & n \ge K \end{cases} \tag{6.8}$$

where $\bar{n}\rho < 1$

$$\rho = \frac{\gamma}{\mu C} \tag{6.9}$$

$$P_0 = \left[\sum_{n=0}^{K-1} \frac{(K\bar{n}\rho)^n}{n!} + \frac{(K\bar{n}\rho)^K}{(1 - \bar{n}\rho)K!} \right]^{-1} \tag{6.10}$$

$$\bar{n} = \frac{N+1}{N} \sum_{r=1}^{N} \frac{1}{1 - (1/K) \sum_{s \in S} \theta^{sr}} \tag{6.11}$$

$\theta = e^{2\pi j/(N+1)} = (N+1)$th primitive root of unity   (6.12)
S = set of integers which correspond to positions of non-zero elements of first row of P

PROOF: We first show that the nodes in the K-connected net obey the conditions of the birth and death process examined earlier. We do this by showing that the birth and death coefficients are independent of time. Specifically, let

Q_1 = event that message on channel connecting node j to node k completes its transmission to node k in arbitrary time interval $(t, t + dt)$

Q_2 = event that channel connecting node j to node k is being used

Clearly, $\Pr[Q_1, Q_2]$ is a component of the death coefficient for node j and a component of the birth coefficient for node k; owing to the symmetry of the traffic matrix and the network topology, these birth and death coefficients are representative of those throughout the net. Now,

$$\Pr[Q_1, Q_2] = \Pr[Q_1|Q_2]\Pr[Q_2]$$

and owing to the independent exponential message lengths,[1]

$$\Pr[Q_1|Q_2] = \frac{\mu C}{K(N+1)}\, dt$$

Also, if we define

$$P_n = \Pr[n \text{ messages in node } j \text{ (say)}]$$

then

$$\Pr[Q_2] = 1 - P_0 - P_1 \frac{K-1}{K} - P_2 \frac{K-2}{K}$$
$$- \cdots - P_{K-1} \frac{1}{K} \quad (D.20)$$

Equation (D.20) is of the form $1 - \Pr[\text{channel connecting node } j \text{ to node } k \text{ is idle}]$. The subscripts j and k disappear, owing to the symmetry of all channels. Note that we have invoked the corollary of Theorem D.1 in obtaining the coefficient for P_n in Eq. (D.20). Rewriting Eq. (D.20), we obtain

$$\Pr[Q_2] = 1 - \sum_{n=0}^{K-1} P_n \frac{K-n}{K} \quad (D.21)$$

Now, we recognize from Theorem 4.1 that

$$\rho_j = 1 - \sum_{n=0}^{\infty} \frac{\bar{C}_n}{C_j} P_n$$

---

[1] Note that we make use of the independence assumption here.

where C_j is the sum of the capacities of all channels leaving node j; because of the symmetry, $C_j = C/(N + 1)$. But, in our case,

$$\bar{C}_n = \begin{cases} \dfrac{K - n}{K} C_j & n \le K \\ 0 & n \ge K \end{cases}$$

Thus Eq. (D.21) becomes

$$P_r[Q_2] = \rho_j = \frac{\Gamma_j}{\mu C_j}$$

where Γ_j is the total arrival rate of messages to node j (from both internal channels and external sources). Owing to symmetry again, $\Gamma_j = \Gamma$. Thus

$$P_r[Q_1, Q_2] = \frac{\mu C}{K(N + 1)} \, dt \, \frac{\Gamma}{\mu C/(N + 1)} = \frac{\Gamma}{K} \, dt \qquad \text{(D.22)}$$

Equation (D.22) states that the interdeparture times of messages are Poisson in nature,[1] at a mean rate of Γ/K; this also implies that the birth and death coefficients for all nodes in the net are independent of time.

We now proceed to evaluate the parameter Γ, which represents the average arrival rate of messages to each node (from both external and internal sources). Each time a new message enters the net from an external source, it brings with it the number of steps it will take before being received at its destination. We may think of this number as being added to the total number x of steps that must be made before all messages currently in the net will be received at their destinations. Similarly, each time a message passes from one node to another, the number x is reduced by 1. Now it is clear that, if the net is in equilibrium, the time derivative of the average value $\bar{x}$ of x (that is, $d\bar{x}/dt$) must be zero. The average rate at which steps are created is equal to the product of the average arrival rate γ of messages to the system and the average number of steps $\bar{n}$ that each message will take. Steps are destroyed at a rate equal to the average number of messages completing transmission per unit time, that is, $\Gamma(N + 1)$. Thus,

$$\gamma \bar{n} = \Gamma(N + 1)$$

or

$$\Gamma = \frac{\gamma \bar{n}}{N + 1}$$

Furthermore, owing to the simple form of the traffic matrix and the

---

[1] This result can be shown to hold for more general nets as well.

circulant form of the network topology and the random routing procedure, the conditions for Theorem 6.1 are satisfied; accordingly, $\bar{n}$ is calculated from Eq. (6.6). Thus, the time-independent birth and death coefficients for a node are, respectively,

$$b_n = \Gamma = \frac{\gamma \bar{n}}{N + 1}$$

$$d_n = \begin{cases} \dfrac{n\mu C}{K(N + 1)} & n \leq K \\[2mm] \dfrac{\mu C}{N + 1} & n \geq K \end{cases}$$

We apply these coefficients to Eq. (A.1) and obtain

$$P_n = \begin{cases} \dfrac{P_0(\bar{n}\rho)^n K^n}{n!} & n \leq K \\[3mm] \dfrac{P_0(\bar{n}\rho)^n K^K}{K!} & n \geq K \end{cases}$$

where $\bar{n}\rho < 1$, and ρ, P_0, and $\bar{n}$ are given in Eqs. (6.9) to (6.11). This completes the proof of Theorem 6.2.

D.6   Theorem 6.3 and Its Proof

Theorem 6.3

For the K-connected net,

$$T = \frac{(N + 1)K\bar{n}}{\mu C} + \frac{\bar{n}(N + 1)}{\mu C(1 - \bar{n}\rho)} \frac{1}{(1 - \bar{n}\rho)S_K + 1} \qquad (6.13)$$

where

$$S_K = \sum_{n=0}^{K-1} \frac{(K\bar{n}\rho)^{n-K} K!}{n!} \qquad (6.14)$$

and ρ and $\bar{n}$ are as defined in Theorem 6.2.

PROOF: We observe that Eq. (6.8) is of the same form as Eq. (A.9) with $N = K$, C replaced with $C/(N + 1)$, and ρ replaced with $\bar{n}\rho$. Thus, the expected time T_0 that a message spends in each node of the net is, by Eq. (A.14),

$$T_0 = \frac{K(N + 1)}{\mu C} + \frac{P(\geq K)}{(1 - \bar{n}\rho)\mu C/(N + 1)}$$

Substituting for $P\ (\geq K)$ and rearranging terms, we get

$$T_0 = \frac{K(N+1)}{\mu C} + \frac{N+1}{\mu C(1-\bar{n}\rho)}\frac{1}{(1-\bar{n}\rho)S_K+1}$$

where S_K is as given in Eq. (6.14).   We now recognize that, for the K-connected net (as defined in Sec. 6.4), Eq. (6.1) must hold exactly (i.e., the expected delay T_0 is the same for all nodes), and so $T = \bar{n}T_0$, which proves Theorem 6.3.

An Operational Description of the Simulation Program

The program is designed to simulate the operation of a wide variety of communication nets such as described in Chap. 1. It was written[1] by the author for the TX-2 (a large-scale high-speed digital computer at Lincoln Laboratory) [52]. The program requires the following specifications for the net it is to simulate:

1. The number of nodes N
2. A topological description—specifically, the capacities of all N^2 channels (many of which may be of zero capacity)
3. The total average number of messages γ entering the net per second
4. The average message lengths[2] $1/\mu$
5. The traffic matrix (whose ij entry represents the *relative* traffic with origin at node i and destination at node j)
6. The number of priority classes and the relative arrival rates of messages from each class
7. A set of lists which describes the routing procedure to be used (see Sec. 7.1)
8. A stopping parameter—namely, the total number of messages to be transmitted through the net

[1] The program consists of approximately 1,800 machine instructions.
[2] All message lengths are chosen from the same exponential distribution.

The maximum number of nodes N that can be accommodated is 36 (this can be extended with some minor program changes). The program, at present, generates interarrival times and message lengths which are exponentially distributed with means $1/\gamma$ and $1/\mu$, respectively. These distributions are obtained either from a built-in radioactive random-number generator or from a psuedo-random-number generator in the program. The restriction to exponential distributions can be removed, and more general distributions may be included with some simple program changes.

The program operates as a *differential event simulator*. By this we mean that the program does not run on synchronous time, but rather time is immediately updated to the time of occurrence of the next random event. Extensive use is made of the tied-list concept, wherein each memory location gives the address of the next memory location in the list. While running, the program generates messages (which are defined by their origins, destinations, arrival times to the network, lengths, and priorities) according to the items specified in the list above, routes these messages through the net (placing them on queues when necessary and obeying the imposed queue discipline), and finally delivers them to their destinations. During this process, the program gathers the statistics called for (see below). Furthermore, the simulation is under the control of the console operator at all times. For example, he may stop the program at any time, observe the accumulated data, and then continue. He may also observe the dynamic operation of the net while it is running. For this last purpose, he uses the console oscilloscope which displays information for visual or photographic consumption. Figure E.1 illustrates a typical display as it might be observed during the simulation. The number in the upper left-hand corner (4453) is the total number of messages generated so far. The three horizontal lines below that indicate certain aspects of the number of messages in the system:[1] the lowest of these displays represents $n(t)$, the current number of messages in the system (in queues and in transmission channels); the middle line represents a geometric average of $n(t)$, that is, a short-term average of $n(t)$; the upper line is the true arithmetic average of $n(t)$. The lower half of the display shows the current number of messages in each node as a vertical line; the node numbers run from left to right. The number displayed to the left of these lines serves to calibrate the horizontal reticle (shown dotted); in this case, the upper dotted line represents 20 messages. Both calibrations are subject to change by continuously

---

[1] The number immediately above these three lines serves to calibrate the vertical reticle (shown dotted), which is superimposed over the three lines. In this case, the full range of the scale is 100 messages.

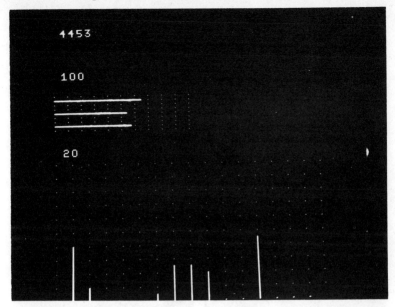

Fig. E.1.   *Photograph of the on-line display.*

variable controls on the console.   In actual operation, this is a dynamic display, wherein the lengths of the solid lines are continuously changing.   This gives the operator a view of the live simulation.   Of course, the display can be suppressed at will by the operator.

For a 13-node net and 10,000 messages, the total running time for the program is of the order of 2 min.   At the termination of the run, a number of gathered statistics may be displayed on the oscilloscope. In fact, this was chosen as the sole output for the program, the idea being that the desired curves could easily be photographed, and the few desired numbers recorded, from the visual display.   Figure E.2a to g illustrates the results of a typical run.   Below each part of the figure is a short description of the contents of the photograph.

In all of the illustrations, the number displayed in the upper left-hand corner represents the height of the maximum data point in the figure.   All displays are automatically scaled so that the maximum height lies between 2^n and 2^{n-1}, where the highest scale line (all scale lines are shown dotted) has height 2^n.   Thus, the reading of quantitative data from these displays is simplified.   In the display of a histogram, the mean value of the distribution represented by the histogram is displayed as a number at the top center of the figure.[1]

---

[1] The number displayed in the center left region of the figure represents the number of time units which are grouped into each bar on the histogram.

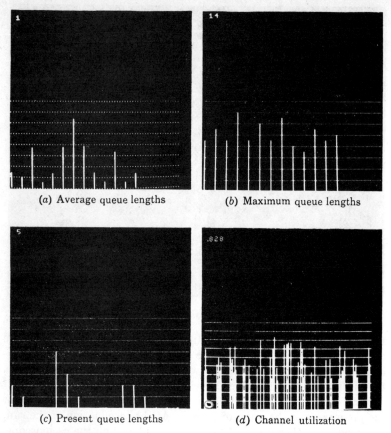

(a) Average queue lengths (b) Maximum queue lengths

(c) Present queue lengths (d) Channel utilization

Fig. E.2. *Statistical output display for a typical simulation run.*

A novel feature is included in the simulation which greatly facili-
tates the experimental procedure. This feature allows automatic
calculation of channel capacities based on the traffic carried by each
channel in the previous run. For example, the channel capacity
assignment described by Eq. (4.7) requires knowledge of the λ_i; the
determination of the λ_i can be a rather tedious calculation for fixed
and especially alternate routing procedures. To assign the C_i automatically, one need merely run the simulation with *any* C_i assignment
initially; at the end of this run, each channel will have carried a particular number of messages. Based on this data, the program can
then be made to calculate a new set of C_i as specified by Eq. (4.7).
Of course, the total capacity distributed through the net remains fixed
during this calculation. If necessary, this procedure may be repeated

(e) Histogram of generated
message lengths

Fig. E.2. (*cont.*) *Statistical output display for a typical simulation run.*

(f) Histogram of generated
inter-arrival times

(g) Histogram of total
message delay

more than once, each time obtaining a closer approximation to the desired assignment (usually two or three iterations are sufficient); in order to decide when the correct assignment has been reached, one need merely compare the assignments for two successive runs (this information is available upon request as a printed list from a high-speed

printer). Furthermore, the program is designed to follow this procedure for other forms of capacity assignment of interest.   Specifically,
the following assignment is included:

$$C_i = \frac{\lambda_i}{\lambda} C \qquad i = 1, 2, \ldots, N^2 \qquad (E.1)$$

where
$$\lambda = \sum_{i=1}^{N^2} \lambda_i$$

and C is the total capacity assigned to the net.   This form of capacity
assignment allocates capacity in direct proportion to the traffic carried.
If desired, the program will assign the same capacity to each channel.[1]

In summary, then, the program is equipped to simulate a large class
of communication nets with different topologies, routing procedures,
priority disciplines, and traffic loads.   It runs quickly, generates messages automatically, and allows visual monitoring of the dynamic operation of the network; upon termination of the run, it visually displays
a number of pertinent statistical distributions with quantitative scales
included.   Moreover, upon request, the program automatically redis-
tributes the total capacity in the network according to one of three
allocation formulas (square root, proportional, and identical capacity
assignment).   This last feature allows on-line experimentation and
optimization of the network parameters.

[1] See footnote on page 108.

appendix **F**

Alternate Routing Theorems and Their Proofs

F.1   Theorem 7.1 and Its Proof

Theorem 7.1

Consider any node (say node n_1) in the network described in Sec. 7.3 from which there are two alternative paths of the same length, both paths leading to the same node (say node n_2). The message delay can always be reduced by omitting one of these paths as an alternative route for messages traveling from n_1 to n_2.

PROOF: Assume that the number of nodes in the net is N. Further, assume that the constant length for both alternative paths between nodes n_1 and n_2 is L (that is, a message will be transmitted over L channels successively in traveling from n_1 to n_2). Let us label those channels included in the first path with the subscripts a_i and those included in the second path with the subscripts b_i ($i = 1, 2, \ldots, L$). Let

λ_{a_i} = average total number of messages per second transmitted over ith channel of path 1

λ_{b_i} = average total number of messages per second transmitted over ith channel of path 2

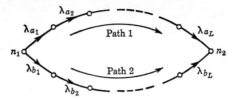

Fig. F.1. *Two equal-length alternative paths.*

λ_1 = average number of messages per second routed from n_1 to n_2 over path 1

λ_2 = average number of messages per second routed from n_1 to n_2 over path 2

(See Fig. F.1, which shows the two paths extracted from a large net.) Further, let the subscript j range over all possible N^2 channels. Since we assume the network traffic to be Poisson in nature, we recognize that, given the set of λ_j, the assignment of C_j which minimizes the average message delay T is set forth by Theorem 4.5. Thus, no matter which routing procedure we use, as long as we maintain independent Poisson traffic, we obtain the minimum message delay by using the square root capacity assignment.

Consider the following change in the traffic pattern:

$$\lambda_1' = \lambda_1 + \alpha$$
$$\lambda_2' = \lambda_2 - \alpha$$

We recognize that

$$\lambda_{a_i}' = \lambda_{a_i} + \alpha$$

and

$$\lambda_{b_i}' = \lambda_{b_i} - \alpha$$

where, of course,

$$-\lambda_1 \leq \alpha \leq \lambda_2 \tag{F.1}$$

Using the square root channel capacity assignment, we find[1] that, for this new traffic pattern (or routing procedure),

$$T' = \frac{\bar{n} \left(\sum_{j=1}^{N^2} \sqrt{\lambda_j'} \right)^2}{\mu \lambda C (1 - \bar{n}\rho)}$$

We now inquire as to the value of α which minimizes T' subject to the constraint in Eq. (F.1). Clearly, minimizing T' with respect to α is the same as minimizing

$$\sum_{j=1}^{N^2} \sqrt{\lambda_j'}$$

[1] See Eq. (4.18).

Furthermore, the only terms in this sum which depend upon α are those for which $j = a_i$ or b_i. Thus, we find that the following equation defines the function we wish to minimize:

$$F = \sum_{i=1}^{L} (\sqrt{\lambda'_{a_i}} + \sqrt{\lambda'_{b_i}}) \qquad (F.2)$$

or

$$F = \sum_{i=1}^{L} (\sqrt{\lambda_{a_i} + \alpha} + \sqrt{\lambda_{b_i} - \alpha})$$

Differentiating, we find that the slope of F is

$$\frac{dF}{d\alpha} = \sum_{i=1}^{L} \left(\frac{1}{2\sqrt{\lambda_{a_i} + \alpha}} - \frac{1}{2\sqrt{\lambda_{b_i} - \alpha}} \right)$$

It is clear that this slope is a monotonically decreasing function of α. F itself must therefore be a convex function of α; that is, for $0 \le \beta \le 1$ and $\alpha_1 \le \alpha_2$,

$$\beta F(\alpha_1) + (1 - \beta)F(\alpha_2) \le F[\beta\alpha_1 + (1 - \beta)\alpha_2]$$

As a result, the minimum value of F is obtained when α is at one of its extreme values $(-\lambda_1 \text{ or } \lambda_2)$. If $\alpha = -\lambda_1$, then all messages traveling between nodes n_1 and n_2 will be routed along path 2; if $\alpha = \lambda_2$, then all this traffic is routed along path 1. This proves Theorem 7.1, since either arrangement corresponds to a fixed routing procedure.

F.2   Theorem 7.2 and Its Proof

Theorem 7.2

For the system described in Sec. 7.3, a message should accept channel C_i if and only if its position n_i in the queue satisfies the following inequalities:

$$n_i - 1 < \frac{S_{i-1}}{C_i} - 1 \le n_i \qquad i = 2, 3, \ldots, N \quad (7.2)$$

where

$$S_{i-1} = \sum_{j=1}^{i-1} C_j \qquad (7.3)$$

and where

$$n_1 = 1$$

PROOF: Consider a message in the queue which has just been offered channel C_i. If the channel is accepted, the message will take, on the average, $1/\mu C_i$ seconds to leave the system. Let us now assume that the message is in that position[1] n_i in the queue for which it makes no difference to the message's expected delay if it accepts or refuses channel C_i. Define, for $i \geq 2$,

E_i = expected time at which message in position n_i will leave system if it does not accept channel C_i at time 0

Then, clearly, n_i must be such that

$$\frac{1}{\mu C_i} = E_i \qquad i \geq 2 \tag{F.3}$$

Furthermore, define

w_i = expected time for message to move from position n_i to position n_{i-1}

Now, since $1/\mu C_i > 1/\mu C_{i-1}$, it is clear from Eq. (F.3) and the definition of n_i that $n_1 < n_2 < \cdots < n_N$. This implies that if channel C_i is offered to the message in position n_i, then channels C_j $(j < i)$ must all be busy (i.e., in the process of transmitting other messages). In such a case, the expected time for our message to move up one position in the queue (i.e., the expected time for any one of these $i - 1$ channels to empty) must be $1/\mu S_{i-1}$ [see Eq. (7.3) for S_{i-1}]. Thus,

$$w_i = \frac{n_i - n_{i-1}}{\mu S_{i-1}} \tag{F.4}$$

Let

p_i = $\Pr$[channel C_i empties before channels C_{i-1}, C_{i-2}, . . . , C_1, given that channels C_i, C_{i-1}, . . . , C_1 are all busy]

Owing to the exponential message lengths,

$$p_i = \frac{C_i}{S_i} \tag{F.5}$$

and

$$1 - p_i = \frac{S_{i-1}}{S_i} \tag{F.6}$$

Now, after our message has reached position n_{i-1}, channel C_{i-1} will empty before any other channel C_j $(j < i - 1)$ with probability p_{i-1}, and the message will accept this channel. On the other hand, one of the other channels C_j will empty first with probability $1 - p_{i-1}$, and then this channel will be accepted by some other message (in position n_j) ahead of our message in the queue; in this case, our message will move up to position $n_{i-1} - 1$ and will spend an additional expected

---

[1] For the moment, we consider n_i to be a continuous variable.

time $E_{i-1} - 1/\mu S_{i-2}$ in the system.   From these considerations, we are able to write the following recursion relation:

$$E_i = w_i + p_{i-1} \frac{1}{\mu C_{i-1}} + (1 - p_{i-1})\left(E_{i-1} - \frac{1}{\mu S_{i-2}} \right) \qquad i \geq 2 \quad (F.7)$$

The solution to Eq. (F.7) is

$$E_i = \frac{n_i + k}{\mu S_{i-1}} \qquad i \geq 2 \qquad\qquad (F.8)$$

where k is some constant yet to be determined.   This solution is easily verified by substituting into Eq. (F.7) and referring to Eqs. (F.4) to (F.6).   Referring back to Eq. (F.3), we now see that

$$n_i = \frac{S_{i-1}}{C_i} - k \qquad i \geq 2 \qquad\qquad (F.9)$$

To evaluate k, we recognize that n_2 must be such that

$$\frac{1}{\mu C_2} = \frac{n_2}{\mu C_1} + \frac{1}{\mu C_1}$$

or

$$n_2 = \frac{C_1}{C_2} - 1$$

and so $k = 1$.

We now recall that n_i must be an integer, so the condition expressed by Eq. (F.9) becomes

$$n_{i-1} < \frac{S_{i-1}}{C_i} - 1 \leq n_i \qquad i \geq 2$$

For $i = 1$, obviously $n_1 = 1$.   This completes the proof of Theorem 7.2.

Bibliography

1. Ford, L. R., Jr., and D. R. Fulkerson: "Flows in Networks," Princeton University Press, Princeton, N.J., 1962.
2. Brockmeyer, E., H. L. Halstrom, and A. Jensen: The Life and Works of A. K. Erlang, *Danish Acad. Tech. Sci.*, no. 2 (1948).
3. Molina, E. C.: Application of the Theory of Probability to Telephone Trunking Problems, *Bell System Tech. J.*, **6**:461–494 (1927).
4. O'Dell, G. F.: Theoretical Principles of the Traffic Capacity of Automatic Switches, *P.O. Elec. Engrs. J.*, **13**:209–223 (1920).
5. Molina, E. C.: The Theory of Probabilities Applied to Telephone Trunking Problems, *Bell System Tech. J.*, **1**(2):69–81 (1922).
6. O'Dell, G. F.: An Outline of the Trunking Aspect of Automatic Telephony, *J. Inst. Elec. Engrs. (London)*, **65**:185–222 (1927).
7. Fry, T. C.: "Probability and Its Engineering Uses," D. Van Nostrand Company, Inc., Princeton, N.J., 1928.
8. Feller, W.: Die Grundlagen der Volterraschen Theorie des Kampfes ums Dasein in wahrscheinlichkeitstheoretischer Behandlung, *Acta Biotheoret.*, **5**:11–40 (1939).
9. Kendall, D. G.: Some Problems in the Theory of Queues, *J. Roy. Statist. Soc.*, Series B, **13**:151–185 (1951).
10. Kendall, D. G.: Stochastic Processes Occurring in the Theory of Queues and Their Analysis by Means of the Imbedded Markov Chain, *Ann. Math. Statistics*, **24**:338–354 (1953).
11. Lindley, D. V.: Theory of Queues with a Single Server, *Proc. Cambridge Phil. Soc.*, **48**(2):277–289 (1952).
12. Palm, C.: Intensity Fluctuations in Telephone Traffic, *Ericsson Tech.*, **1**(44):1–189 (1943).
13. Riordan, J.: Telephone Traffic Time Averages, *Bell System Tech. J.*, **30**:1129–1144 (1951).

14. Morse, P. M.: Stochastic Properties of Waiting Lines, *Operations Res.*, **3**:255–261 (1955).
15. Burke, P. J.: The Output of a Qucueing System, *Operations Res.*, **4**:699–704 (1956).
16. Foster, F. G.: A Unified Theory for Stock, Storage and Queue Control, *Operations Res. Quart.*, **10**:121–130 (1959).
17. Little, J. D. C.: A Proof for the Queueing Formula $L = \lambda W$, *Operations Res.*, **9**:383–387 (1961).
18. Feller, W.: "An Introduction to Probability Theory and Its Applications," John Wiley & Sons, Inc., New York, 1950.
19. Pollaczek, F.: Problèmes stochastiques posés par le phénomène de formation d'une queue d'attente à un guichet et par des phénomènes apparentés, *Mém. Sci. Math. (Paris)*, no. 136 (1957).
20. Morse, P. M.: "Queues, Inventories, and Maintenance," John Wiley & Sons, Inc., New York, 1958.
21. Khinchine, A. Ja.: "Mathematical Methods in the Theory of Queueing," translated by D. M. Andrews and M. H. Quenouille, Charles Griffin & Company, Ltd., London, 1960.
22. Syski, R.: "Introduction to Congestion in Telephone Systems," Oliver & Boyd Ltd., Edinburgh and London, 1960.
23. Saaty, T. L.: "Elements of Queueing Theory with Applications," McGraw-Hill Book Company, New York, 1961.
24. Cox, D. R., and W. L. Smith: "Queues," Methuen & Co., Ltd., London, and John Wiley & Sons, Inc., New York, 1961.
25. Riordan, J.: "Stochastic Service Systems," John Wiley & Sons, Inc., New York, 1961.
26. Elias, P., A. Feinstein, and C. E. Shannon: A Note on the Maximum Flow Through a Network, *IRE Trans. Inform. Theory*, **IT-2**:117–119 (1956).
27. Gomory, R. E., and T. C. Hu: Multi-terminal Network Flows, *J. Soc. Ind. Appl. Math.*, **9**(4):551–570 (1961).
28. Gomory, R. E., and T. C. Hu: An Application of Generalized Linear Programming to Network Flows, *J. Soc. Ind. Appl. Math.*, **10**(2):260–283, 1962.
29. Chien, R. T.: Synthesis of a Communication Net, *IBM J. Res. Develop.*, **4**:311–320 (1960).
30. Mayeda, W.: Terminal and Branch Capacity Matrices of a Communication Net, *IRE Trans. on Circuit Theory*, **7**:251–269 (1960).
31. Gomory, R. E., and T. C. Hu: "Multi-commodity Network Flows," I.B.M. research report RC-865, January 16, 1963.
32. Jewell, W. S.: "Optimal Flow Through a Network," M.I.T. doctorate thesis (Electrical Engineering Department), 1958.

33. Robacker, J. T.: "Concerning Multi-commodity Flows," Rand report RM-1799, 1956.
34. Hakimi, S. L.: On Simultaneous Flows in a Communication Network, *IRE Trans. on Circuit Theory*, **CT-9**(2):169–175 (1962).
35. Prihar, Z.: Topological Properties of Telecommunication Networks, *Proc. IRE*, 44:927–933 (1956).
36. Hunt, G. C.: Sequential Arrays of Waiting Lines, *Operations Res.*, 4:674–683 (1956).
37. Jackson, J. R.: Networks of Waiting Lines, *Operations Res.*, 5:518–521 (1957).
38. Shannon, C. E.: Memory Requirements in a Telephone Exchange, *Bell System Tech. J.*, 29:343–349 (1950).
39. Moran, P. A. P.: "The Theory of Storage," Methuen & Co., Ltd., London, and John Wiley & Sons, Inc., New York, 1959.
40. Prosser, R. T.: Routing Procedures in Communications Networks, part I: Random Procedures, *IRE Trans. on Commun. Systems*, **CS-10**(4):322–329 (1962).
41. Prosser, R. T.: Routing Procedures in Communications Networks, part II: Directory Procedures, *IRE Trans. on Commun. Systems*, **CS-10**(4):329–335 (1962).
42. Shannon, C. E., and W. Weaver: "The Mathematical Theory of Communication," The University of Illinois Press, Urbana, Ill., 1949.
43. Vernam, G. S.: Automatic Telegraph Switching System Plan 55A, *Western Union Tech. Rev.*, **12**(2):37–50 (1958).
44. Zipf, G. K.: "Human Behavior and the Principle of Least Effort," Addison-Wesley Publishing Company, Inc., Reading, Mass., 1949.
45. Whitten, C. A.: "Air-line Distances between Cities in the United States," U.S. Coast and Geodetic Survey special publication 238, 1947.
46. Hardy, G. H., J. E. Littlewood, and G. Polya: "Inequalities," Cambridge University Press, New York, 1959.
47. Phipps, T. E., Jr.: Machine Repair as a Priority Waiting-line Problem, *Operations Res.*, 4:76–85 (1956). (Comments by W. R. Van Voorhis, *ibid.*, p. 86.)
48. Cobham, A.: Priority Assignments in Waiting Line Problems, *Operations Res.*, 2:70–76 (1954); A Correction, *Operations Res.*, 3:547 (1955). See also J. L. Holley, Waiting Line Subject to Priorities, *Operations Res.*, 2:341–343 (1954).
49. White, H., and L. S. Christie: Queueing with Pre-emptive Priorities or with Breakdown, *Operations Res.*, 6:79–95 (1958).
50. Jackson, J. R.: Some Problems in Queueing with Dynamic Priorities, *Naval Res. Logistics Quart.*, 7:235–249 (1960).

51. Jackson, J. R.: Waiting Time Distributions for Queues with Dynamic Priorities, *Naval Res. Logistics Quart.*, **9**:31–36 (1962).
52. Frankovich, J. M., and H. P. Peterson: A Functional Description of the Lincoln TX-2 Computer, *Proc. Western Joint Computer Conf.*, pp. 146–155 (1957).
53. Hildebrand, F. B.: "Methods of Applied Mathematics," Prentice-Hall, Inc., Englewood Cliffs, N.J., 1958.
54. Kemperman, J. H. B.: "The Passage Problem for a Stationary Markov Chain," The University of Chicago Press, Chicago, 1961.

Index

A CATALOGUE OF SELECTED
DOVER SCIENCE BOOKS

Physics: The Pioneer Science, Lloyd W. Taylor. Very thorough non-mathematical survey of physics in a historical framework which shows development of ideas. Easily followed by laymen; used in dozens of schools and colleges for survey courses. Richly illustrated. Volume 1: Heat, sound, mechanics. Volume 2: Light, electricity. Total of 763 illustrations. Total of cvi + 847pp.
                    60565-5, 60566-3 Two volumes, Paperbound 5.50

THE RISE OF THE NEW PHYSICS, A. d'Abro. Most thorough explanation in print of central core of mathematical physics, both classical and modern, from Newton to Dirac and Heisenberg. Both history and exposition: philosophy of science, causality, explanations of higher mathematics, analytical mechanics, electromagnetism, thermodynamics, phase rule, special and general relativity, matrices. No higher mathematics needed to follow exposition, though treatment is elementary to intermediate in level. Recommended to serious student who wishes verbal understanding. 97 illustrations. Total of ix + 982pp.
                    20003-5, 20004-3 Two volumes, Paperbound $6.00

INTRODUCTION TO CHEMICAL PHYSICS, John C. Slater. A work intended to bridge the gap between chemistry and physics. Text divided into three parts: Thermodynamics, Statistical Mechanics, and Kinetic Theory; Gases, Liquids and Solids; and Atoms, Molecules and the Structure of Matter, which form the basis of the approach. Level is advanced undergraduate to graduate, but theoretical physics held to minimum. 40 tables, 118 figures. xiv + 522pp.
                    62562-1 Paperbound $4.00

BASIC THEORIES OF PHYSICS, Peter C. Bergmann. Critical examination of important topics in classical and modern physics. Exceptionally useful in examining conceptual framework and methodology used in construction of theory. Excellent supplement to any course, textbook. Relatively advanced.
Volume 1. Heat and Quanta. Kinetic hypothesis, physics and statistics, stationary ensembles, thermodynamics, early quantum theories, atomic spectra, probability waves, quantization in wave mechanics, approximation methods, abstract quantum theory. 8 figures. x + 300pp.   60968-5 Paperbound $2.50
Volume 2. Mechanics and Electrodynamics. Classical mechanics, electro- and magnetostatics, electromagnetic induction, field waves, special relativity, waves, etc. 16 figures, viii + 260pp.   60969-3 Paperbound $2.75

FOUNDATIONS OF PHYSICS, Robert Bruce Lindsay and Henry Margenau. Methods and concepts at the heart of physics (space and time, mechanics, probability, statistics, relativity, quantum theory) explained in a text that bridges gap between semi-popular and rigorous introductions. Elementary calculus assumed. "Thorough and yet not over-detailed," *Nature*. 35 figures. xviii + 537 pp.
                    60377-6 Paperbound $3.50

A CATALOGUE OF SELECTED
DOVER SCIENCE BOOKS

FUNDAMENTAL FORMULAS OF PHYSICS, edited by Donald H. Menzel. Most useful reference and study work, ranges from simplest to most highly sophisticated operations. Individual chapters, with full texts explaining formulae, prepared by leading authorities cover basic mathematical formulas, statistics, nomograms, physical constants, classical mechanics, special theory of relativity, general theory of relativity, hydrodynamics and aerodynamics, boundary value problems in mathematical physics, heat and thermodynamics, statistical mechanics, kinetic theory of gases, viscosity, thermal conduction, electromagnetism, electronics, acoustics, geometrical optics, physical optics, electron optics, molecular spectra, atomic spectra, quantum mechanics, nuclear theory, cosmic rays and high energy phenomena, particle accelerators, solid state, magnetism, etc. Special chapters also cover physical chemistry, astrophysics, celestian mechanics, meteorology, and biophysics. Indispensable part of library of every scientist. Total of xli + 787pp.
60595-7, 60596-5 Two volumes, Paperbound $6.00

INTRODUCTION TO EXPERIMENTAL PHYSICS, William B. Fretter. Detailed coverage of techniques and equipment: measurements, vacuum tubes, pulse circuits, rectifiers, oscillators, magnet design, particle counters, nuclear emulsions, cloud chambers, accelerators, spectroscopy, magnetic resonance, x-ray diffraction, low temperature, etc. One of few books to cover laboratory hazards, design of exploratory experiments, measurements. 298 figures. xii + 349pp.
(EBE) 61890-0 Paperbound $3.00

CONCEPTS AND METHODS OF THEORETICAL PHYSICS, Robert Bruce Lindsay. Introduction to methods of theoretical physics, emphasizing development of physical concepts and analysis of methods. Part I proceeds from single particle to collections of particles to statistical method. Part II covers application of field concept to material and non-material media. Numerous exercises and examples. 76 illustrations. x + 515pp.
62354-8 Paperbound $4.00

AN ELEMENTARY TREATISE ON THEORETICAL MECHANICS, Sir James Jeans. Great scientific expositor in remarkably clear presentation of basic classical material: rest, motion, forces acting on particle, statics, motion of particle under variable force, motion of rigid bodies, coordinates, etc. Emphasizes explanation of fundamental physical principles rather than mathematics or applications. Hundreds of problems worked in text. 156 figures. x + 364pp.   61839-0 Paperbound $2.75

THEORETICAL MECHANICS: AN INTRODUCTION TO MATHEMATICAL PHYSICS, Joseph S. Ames and Francis D. Murnaghan. Mathematically rigorous introduction to vector and tensor methods, dynamics, harmonic vibrations, gyroscopic theory, principle of least constraint, Lorentz-Einstein transformation. 159 problems; many fully-worked examples. 39 figures. ix + 462pp.   60461-6 Paperbound $3.50

THE PRINCIPLE OF RELATIVITY, Albert Einstein, Hendrick A. Lorentz, Hermann Minkowski and Hermann Weyl. Eleven original papers on the special and general theory of relativity, all unabridged. Seven papers by Einstein, two by Lorentz, one each by Minkowski and Weyl. "A thrill to read again the original papers by these giants," *School Science and Mathematics*. Translated by W. Perret and G. B. Jeffery. Notes by A. Sommerfeld. 7 diagrams. viii + 216pp.
60081-5 Paperbound $2.25

EINSTEIN'S THEORY OF RELATIVITY, Max Born. Relativity theory analyzed, explained for intelligent layman or student with some physical, mathematical background. Includes Lorentz, Minkowski, and others. Excellent verbal account for teachers. Generally considered the finest non-technical account. vii + 376pp.
60769-0 Paperbound $2.75

PHYSICAL PRINCIPLES OF THE QUANTUM THEORY, Werner Heisenberg. Nobel Laureate discusses quantum theory, uncertainty principle, wave mechanics, work of Dirac, Schroedinger, Compton, Wilson, Einstein, etc. Middle, non-mathematical level for physicist, chemist not specializing in quantum; mathematical appendix for specialists. Translated by C. Eckart and F. Hoyt. 19 figures. viii + 184pp.
60113-7 Paperbound $2.00

PRINCIPLES OF QUANTUM MECHANICS, William V. Houston. For student with working knowledge of elementary mathematical physics; uses Schroedinger's wave mechanics. Evidence for quantum theory, postulates of quantum mechanics, applications in spectroscopy, collision problems, electrons, similar topics. 21 figures. 288pp.
60524-8 Paperbound $3.00

ATOMIC SPECTRA AND ATOMIC STRUCTURE, Gerhard Herzberg. One of the best introductions to atomic spectra and their relationship to structure; especially suited to specialists in other fields who require a comprehensive basic knowledge. Treatment is physical rather than mathematical. 2nd edition. Translated by J. W. T. Spinks. 80 illustrations. xiv + 257pp.
60115-3 Paperbound $2.00

ATOMIC PHYSICS: AN ATOMIC DESCRIPTION OF PHYSICAL PHENOMENA, Gaylord P. Harnwell and William E. Stephens. One of the best introductions to modern quantum ideas. Emphasis on the extension of classical physics into the realms of atomic phenomena and the evolution of quantum concepts. 156 problems. 173 figures and tables. xi + 401pp.
61584-7 Paperbound $3.00

ATOMS, MOLECULES AND QUANTA, Arthur E. Ruark and Harold C. Urey. 1964 edition of work that has been a favorite of students and teachers for 30 years. Origins and major experimental data of quantum theory, development of concepts of atomic and molecular structure prior to new mechanics, laws and basic ideas of quantum mechanics, wave mechanics, matrix mechanics, general theory of quantum dynamics. Very thorough, lucid presentation for advanced students. 230 figures. Total of xxiii + 810pp.
61106-X, 61107-8 Two volumes, Paperbound $6.00

INVESTIGATIONS ON THE THEORY OF THE BROWNIAN MOVEMENT, Albert Einstein. Five papers (1905-1908) investigating the dynamics of Brownian motion and evolving an elementary theory of interest to mathematicians, chemists and physical scientists. Notes by R. Fürth, the editor, discuss the history of study of Brownian movement, elucidate the text and analyze the significance of the papers. Translated by A. D. Cowper. 3 figures. iv + 122pp.
60304-0 Paperbound $1.50

MATHEMATICAL FOUNDATIONS OF STATISTICAL MECHANICS, A. I. Khinchin. Introduction to modern statistical mechanics: phase space, ergodic problems, theory of probability, central limit theorem, ideal monatomic gas, foundation of thermo-dynamics, dispersion and distribution of sum functions. Provides mathematically rigorous treatment and excellent analytical tools. Translated by George Gamow. viii + 179pp. 60147-1 Paperbound $2.50

INTRODUCTION TO PHYSICAL STATISTICS, Robert B. Lindsay. Elementary prob-ability theory, laws of thermodynamics, classical Maxwell-Boltzmann statistics, classical statistical mechanics, quantum mechanics, other areas of physics that can be studied statistically. Full coverage of methods; basic background theory. ix + 306pp. 61882-X Paperbound $2.75

DIALOGUES CONCERNING TWO NEW SCIENCES, Galileo Galilei. Written near the end of Galileo's life and encompassing 30 years of experiment and thought, these dialogues deal with geometric demonstrations of fracture of solid bodies, cohesion, leverage, speed of light and sound, pendulums, falling bodies, accelerated motion, etc. Translated by Henry Crew and Alfonso de Salvio. Introduction by Antonio Favaro. xxiii + 300pp. 60099-8 Paperbound $2.25

FOUNDATIONS OF SCIENCE: THE PHILOSOPHY OF THEORY AND EXPERIMENT, Norman R. Campbell. Fundamental concepts of science examined on middle level: acceptance of propositions and axioms, presuppositions of scientific thought, scien-tific law, multiplication of probabilities, nature of experiment, application of math-ematics, measurement, numerical laws and theories, error, etc. Stress on physics, but holds for other sciences. "Unreservedly recommended," *Nature* (England). Formerly *Physics: The Elements*. ix + 565pp. 60372-5 Paperbound $4.00

THE PHASE RULE AND ITS APPLICATIONS, Alexander Findlay, A. N. Campbell and N. O. Smith. Findlay's well-known classic, updated (1951). Full standard text and thorough reference, particularly useful for graduate students. Covers chemical phenomena of one, two, three, four and multiple component systems. "Should rank as the standard work in English on the subject," *Nature*. 236 figures. xii + 494pp. 60091-2 Paperbound $3.50

THERMODYNAMICS, Enrico Fermi. A classic of modern science. Clear, organized treatment of systems, first and second laws, entropy, thermodynamic potentials, gaseous reactions, dilute solutions, entropy constant. No math beyond calculus is needed, but readers are assumed to be familiar with fundamentals of thermometry, calorimetry. 22 illustrations. 25 problems. x + 160pp.
60361-X Paperbound $2.00

TREATISE ON THERMODYNAMICS, Max Planck. Classic, still recognized as one of the best introductions to thermodynamics. Based on Planck's original papers, it presents a concise and logical view of the entire field, building physical and chemical laws from basic empirical facts. Planck considers fundamental definitions, first and second principles of thermodynamics, and applications to special states of equilibrium. Numerous worked examples. Translated by Alexander Ogg. 5 figures. xiv + 297pp. 60219-2 Paperbound $2.50

MICROSCOPY FOR CHEMISTS, Harold F. Schaeffer. Thorough text; operation of microscope, optics, photomicrographs, hot stage, polarized light, chemical procedures for organic and inorganic reactions. 32 specific experiments cover specific analyses: industrial, metals, other important subjects. 136 figures. 264pp.
61682-7 Paperbound $2.50

OPTICKS, Sir Isaac Newton. A survey of 18th-century knowledge on all aspects of light as well as a description of Newton's experiments with spectroscopy, colors, lenses, reflection, refraction, theory of waves, etc. in language the layman can follow. Foreword by Albert Einstein. Introduction by Sir Edmund Whittaker. Preface by I. Bernard Cohen. cxxvi + 406pp.
60205-2 Paperbound $4.00

LIGHT: PRINCIPLES AND EXPERIMENTS, George S. Monk. Thorough coverage, for student with background in physics and math, of physical and geometric optics. Also includes 23 experiments on optical systems, instruments, etc. "Probably the best intermediate text on optics in the English language," *Physics Forum*. 275 figures. xi + 489pp.
60341-5 Paperbound $3.50

PHYSICAL OPTICS, Robert W. Wood. A classic in the field, this is a valuable source for students of physical optics and excellent background material for a study of electromagnetic theory. Partial contents: nature and rectilinear propagation of light, reflection from plane and curved surfaces, refraction, absorption and dispersion, origin of spectra, interference, diffraction, polarization, Raman effect, optical properties of metals, resonance radiation and fluorescence of atoms, magneto-optics, electro-optics, thermal radiation. 462 diagrams, 17 plates. xvi + 846pp.
61808-0 Paperbound $4.50

MIRRORS, PRISMS AND LENSES: A TEXTBOOK OF GEOMETRICAL OPTICS, James P. C. Southall. Introductory-level account of modern optical instrument theory, covering unusually wide range: lights and shadows, reflection of light and plane mirrors, refraction, astigmatic lenses, compound systems, aperture and field of optical system, the eye, dispersion and achromatism, rays of finite slope, the microscope, much more. Strong emphasis on earlier, elementary portions of field, utilizing simplest mathematics wherever possible. Problems. 329 figures. xxiv + 806pp.
61234-1 Paperbound $5.00

THE PSYCHOLOGY OF INVENTION IN THE MATHEMATICAL FIELD, Jacques Hadamard. Important French mathematician examines psychological origin of ideas, role of the unconscious, importance of visualization, etc. Based on own experiences and reports by Dalton, Pascal, Descartes, Einstein, Poincaré, Helmholtz, etc. xiii + 145pp.
20107-4 Paperbound $1.50

INTRODUCTION TO CHEMICAL PHYSICS, John C. Slater. A work intended to bridge the gap between chemistry and physics. Text divided into three parts: Thermodynamics, Statistical Mechanics, and Kinetic Theory; Gases, Liquids and Solids; and Atoms, Molecules and the Structure of Matter, which form the basis of the approach. Level is advanced undergraduate to graduate, but theoretical physics held to minimum. 40 tables, 118 figures. xiv + 522pp.
62562-1 Paperbound $4.00

CONTRIBUTIONS TO THE FOUNDING OF THE THEORY OF TRANSFINITE NUMBERS, Georg Cantor. The famous articles of 1895-1897 which founded a new branch of mathematics, translated with 82-page introduction by P. Jourdain. Not only a great classic but still one of the best introductions for the student. ix + 211pp.
60045-9 Paperbound $2.50

ESSAYS ON THE THEORY OF NUMBERS, Richard Dedekind. Two classic essays, on the theory of irrationals, giving an arithmetic and rigorous foundation; and on transfinite numbers and properties of natural numbers. Translated by W. W. Beman. iii + 115pp.
21010-3 Paperbound $1.75

GEOMETRY OF FOUR DIMENSIONS, H. P. Manning. Part verbal, part mathematical development of fourth dimensional geometry. Historical introduction. Detailed treatment is by synthetic method, approaching subject through Euclidean geometry. No knowledge of higher mathematics necessary. 76 figures. ix + 348pp.
60182-X Paperbound $3.00

AN INTRODUCTION TO THE GEOMETRY OF N DIMENSIONS, Duncan M. Y. Sommerville. The only work in English devoted to higher-dimensional geometry. Both metric and projectiv ʿproperties of n-dimensional geometry are covered. Covers fundamental ideas of incidence, parallelism, perpendicularity, angles between linear space, enumerative geometry, analytical geometry, polytopes, analysis situs, hyperspacial figures. 60 diagrams. xvii + 196pp.
60494-2 Paperbound $2.00

THE THEORY OF SOUND, J. W. S. Rayleigh. Still valuable classic by the great Nobel Laureate. Standard compendium summing up previous research and Rayleigh's original contributions. Covers harmonic vibrations, vibrating systems, vibrations of strings, membranes, plates, curved shells, tubes, solid bodies, refraction of plane waves, general equations. New historical introduction and bibliography by R. B. Lindsay, Brown University. 97 figures. lviii + 984pp.
60292-3, 60293-1 Two volumes, Paperbound $6.00

ELECTROMAGNETIC THEORY: A CRITICAL EXAMINATION OF FUNDAMENTALS, Alfred O'Rahilly. Critical analysis and restructuring of the basic theories and ideas of classical electromagnetics. Analysis is carried out through study of the primary treatises of Maxwell, Lorentz, Einstein, Weyl, etc., which established the theory. Expansive reference to and direct quotation from these treatises. Formerly *Electromagnetics*. Total of xvii + 884pp.
60126-9, 60127-7 Two volumes, Paperbound $6.00

ELEMENTARY CONCEPTS OF TOPOLOGY, Paul Alexandroff. Elegent, intuitive approach to topology, from the basic concepts of set-theoretic topology to the concept of Betti groups. Stresses concepts of complex, cycle and homology. Shows how concepts of topology are useful in math and physics. Introduction by David Hilbert. Translated by Alan E. Farley. 25 figures. iv + 57pp.
60747-X Paperbound $1.25

THE ELEMENTS OF NON-EUCLIDEAN GEOMETRY, Duncan M. Y. Sommerville. Presentation of the development of non-Euclidean geometry in logical order, from a fundamental analysis of the concept of parallelism to such advanced topics as inversion, transformations, pseudosphere, geodesic representation, relation between parataxy and parallelism, etc. Knowledge of only high-school algebra and geometry is presupposed. 126 problems, 129 figures. xvi + 274pp.
60460-8 Paperbound $2.50

NON-EUCLIDEAN GEOMETRY: A CRITICAL AND HISTORICAL STUDY OF ITS DEVELOPMENT, Roberto Bonola. Standard survey, clear, penetrating, discussing many systems not usually represented in general studies. Easily followed by non-specialist. Translated by H. Carslaw. Bound in are two most important texts: Bolyai's "The Science of Absolute Space" and Lobachevski's "The Theory of Parallels," translated by G. B. Halsted. Introduction by F. Enriques. 181 diagrams. Total of 431pp.
60027-0 Paperbound $3.00

ELEMENTS OF NUMBER THEORY, Ivan M. Vinogradov. By stressing demonstrations and problems, this modern text can be understood by students without advanced math backgrounds. "A very welcome addition," *Bulletin, American Mathematical Society.* Translated by Saul Kravetz. Over 200 fully-worked problems. 100 numerical exercises. viii + 227pp.
60259-1 Paperbound $2.50

THEORY OF SETS, E. Kamke. Lucid introduction to theory of sets, surveying discoveries of Cantor, Russell, Weierstrass, Zermelo, Bernstein, Dedekind, etc. Knowledge of college algebra is sufficient background. "Exceptionally well written," *School Science and Mathematics.* Translated by Frederick Bagemihl. vii + 144pp.
60141-2 Paperbound $1.75

A TREATISE ON THE DIFFERENTIAL GEOMETRY OF CURVES AND SURFACES, Luther P. Eisenhart. Detailed, concrete introductory treatise on differential geometry, developed from author's graduate courses at Princeton University. Thorough explanation of the geometry of curves and surfaces, concentrating on problems most helpful to students. 683 problems, 30 diagrams. xiv + 474pp.
60667-8 Paperbound $3.50

AN ESSAY ON THE FOUNDATIONS OF GEOMETRY, Bertrand Russell. A mathematical and physical analysis of the place of the a priori in geometric knowledge. Includes critical review of 19th-century work in non-Euclidean geometry as well as illuminating insights of one of the great minds of our time. New foreword by Morris Kline. xx + 201pp.
60233-8 Paperbound $2.50

INTRODUCTION TO THE THEORY OF NUMBERS, Leonard E. Dickson. Thorough, comprehensive approach with adequate coverage of classical literature, yet simple enough for beginners. Divisibility, congruences, quadratic residues, binary quadratic forms, primes, least residues, Fermat's theorem, Gauss's lemma, and other important topics. 249 problems, 1 figure. viii + 183pp.
60342-3 Paperbound $2.00

AN ELEMENTARY INTRODUCTION TO THE THEORY OF PROBABILITY, B. V. Gnedenko and A. Ya. Khinchin. Introduction to facts and principles of probability theory. Extremely thorough within its range. Mathematics employed held to elementary level. Excellent, highly accurate layman's introduction. Translated from the fifth Russian edition by Leo Y. Boron. xii + 130pp.

60155-2 Paperbound $2.00

SELECTED PAPERS ON NOISE AND STOCHASTIC PROCESSES, edited by Nelson Wax. Six papers which serve as an introduction to advanced noise theory and fluctuation phenomena, or as a reference tool for electrical engineers whose work involves noise characteristics, Brownian motion, statistical mechanics. Papers are by Chandrasekhar, Doob, Kac, Ming, Ornstein, Rice, and Uhlenbeck. Exact facsimile of the papers as they appeared in scientific journals. 19 figures. v + 337pp. 6⅛ x 9¼.

60262-1 Paperbound $3.50

STATISTICS MANUAL, Edwin L. Crow, Frances A. Davis and Margaret W. Maxfield. Comprehensive, practical collection of classical and modern methods of making statistical inferences, prepared by U. S. Naval Ordnance Test Station. Formulae, explanations, methods of application are given, with stress on use. Basic knowledge of statistics is assumed. 21 tables, 11 charts, 95 illustrations. xvii + 288pp.

60599-X Paperbound $2.50

MATHEMATICAL FOUNDATIONS OF INFORMATION THEORY, A. I. Khinchin. Comprehensive introduction to work of Shannon, McMillan, Feinstein and Khinchin, placing these investigations on a rigorous mathematical basis. Covers entropy concept in probability theory, uniqueness theorem, Shannon's inequality, ergodic sources, the E property, martingale concept, noise, Feinstein's fundamental lemma, Shanon's first and second theorems. Translated by R. A. Silverman and M. D. Friedman. iii + 120pp.

60434-9 Paperbound $1.75

INTRODUCTION TO SYMBOLIC LOGIC AND ITS APPLICATION, Rudolf Carnap. Clear, comprehensive, rigorous introduction. Analysis of several logical languages. Investigation of applications to physics, mathematics, similar areas. Translated by Wiliam H. Meyer and John Wilkinson. xiv + 214pp.

60453-5 Paperbound $2.50

SYMBOLIC LOGIC, Clarence I. Lewis and Cooper H. Langford. Probably the most cited book in the literature, with much material not otherwise obtainable. Paradoxes, logic of extensions and intensions, converse substitution, matrix system, strict limitations, existence of terms, truth value systems, similar material. vii + 518pp.

60170-6 Paperbound $4.50

VECTOR AND TENSOR ANALYSIS, George E. Hay. Clear introduction; starts with simple definitions, finishes with mastery of oriented Cartesian vectors, Christoffel symbols, solenoidal tensors, and applications. Many worked problems show applications. 66 figures. viii + 193pp.

60109-9 Paperbound $2.50

GUIDE TO THE LITERATURE OF MATHEMATICS AND PHYSICS, INCLUDING RELATED WORKS ON ENGINEERING SCIENCE, Nathan Grier Parke III. This up-to-date guide puts a library catalog at your fingertips. Over 5000 entries in many languages under 120 subject headings, including many recently available Russian works. Citations are as full as possible, and cross-references and suggestions for further investigation are provided. Extensive listing of bibliographical aids. 2nd revised edition. Complete indices. xviii + 436pp.

60447-0 Paperbound $3.00

INTRODUCTION TO ELLIPTIC FUNCTIONS WITH APPLICATIONS, Frank Bowman. Concise, practical introduction, from familiar trigonometric function to Jacobian elliptic functions to applications in electricity and hydrodynamics. Legendre's standard forms for elliptic integrals, conformal representation, etc., fully covered. Requires knowledge of basic principles of differentiation and integration only. 157 problems and examples, 56 figures. 115pp. 60922-7 Paperbound $1.50

THEORY OF FUNCTIONS OF A COMPLEX VARIABLE, A. R. Forsyth. Standard, classic presentation of theory of functions, stressing multiple-valued functions and related topics: theory of multiform and uniform periodic functions, Weierstrass's results with additiontheorem functions. Riemann functions and surfaces, algebraic functions, Schwarz's proof of the existence-theorem, theory of conformal mapping, etc. 125 figures, 1 plate. Total of xxviii + 855pp. $6\frac{1}{8}$ x $9\frac{1}{4}$.

61378-X, 61379-8 Two volumes, Paperbound $6.00

THEORY OF THE INTEGRAL, Stanislaw Saks. Excellent introduction, covering all standard topics: set theory, theory of measure, functions with general properties, and theory of integration emphasizing the Lebesgue integral. Only a minimal background in elementary analysis needed. Translated by L. C. Young. 2nd revised edition. xv + 343pp. 61151-5 Paperbound $3.00

THE THEORY OF FUNCTIONS, *Konrad Knopp. Characterized as "an excellent introduction . . . remarkably readable, concise, clear, rigorous" by the* Journal of the American Statistical Association *college text.*

A COURSE IN MATHEMATICAL ANALYSIS, Edouard Goursat. *The entire "Cours d'analyse" for students with one year of calculus, offering an exceptionally wide range of subject matter on analysis and applied mathematics. Available for the first time in English. Definitive treatment.*

VOLUME I: Applications to geometry, expansion in series, definite integrals, derivatives and differentials. Translated by Earle R. Hedrick. 52 figures. viii + 548pp. 60554-X Paperbound $5.00

VOLUME II, PART I: Functions of a complex variable, conformal representations, doubly periodic functions, natural boundaries, etc. Translated by Earle R. Hedrick and Otto Dunkel. 38 figures. x + 259pp. 60555-8 Paperbound $3.00

VOLUME II, PART II: Differential equations, Cauchy-Lipschitz method, non-linear differential equations, simultaneous equations, etc. Translated by Earle R. Hedrick and Otto Dunkel. 1 figure. viii + 300pp. 60556-6 Paperbound $3.00

VOLUME III, PART I: Variation of solutions, partial differential equations of the second order. Poincaré's theorem, periodic solutions, asymptotic series, wave propagation, Dirichlet's problem in space, Newtonian potential, etc. Translated by Howard G. Bergmann. 15 figures. x + 329pp.   61176-0 Paperbound $3.50

VOLUME III, PART II: Integral equations and calculus of variations: Fredholm's equation, Hilbert-Schmidt theorem, symmetric kernels, Euler's equation, transversals, extreme fields, Weierstrass's theory, etc. Translated by Howard G. Bergmann. Note on Conformal Representation by Paul Montel. 13 figures. xi + 389pp.
61177-9 Paperbound $3.00

ELEMENTARY STATISTICS: WITH APPLICATIONS IN MEDICINE AND THE BIOLOGICAL SCIENCES, Frederick E. Croxton. Presentation of all fundamental techniques and methods of elementary statistics assuming average knowledge of mathematics only. Useful to readers in all fields, but many examples drawn from characteristic data in medicine and biological sciences. vii + 376pp.
60506-X Paperbound $2.50

ELEMENTS OF THE THEORY OF FUNCTIONS. A general background text that explores complex numbers, linear functions, sets and sequences, conformal mapping. Detailed proofs. Translated by Frederick Bagemihl. 140pp.
60154-4 Paperbound $1.50

THEORY OF FUNCTIONS, PART I. Provides full demonstrations, rigorously set forth, of the general foundations of the theory: integral theorems, series, the expansion of analytic functions. Translated by Federick Bagemihl. vii + 146pp.
60156-0 Paperbound $1.50

INTRODUCTION TO THE THEORY OF FOURIER'S SERIES AND INTEGRALS, Horatio S. Carslaw. A basic introduction to the theory of infinite series and integrals, with special reference to Fourier's series and integrals. Based on the classic Riemann integral and dealing with only ordinary functions, this is an important class text. 84 examples. xiii + 368pp.   60048-3 Paperbound $3.00

AN INTRODUCTION TO FOURIER METHODS AND THE LAPLACE TRANSFORMATION, Philip Franklin. Introductory study of theory and applications of Fourier series and Laplace transforms, for engineers, physicists, applied mathematicians, physical science teachers and students. Only a previous knowledge of elementary calculus is assumed. Methods are related to physical problems in heat flow, vibrations, eletcrical transmission, electromagnetic radiation, etc. 828 problems with answers. Formerly *Fourier Methods.* x + 289pp.   60452-7 Paperbound $2.75

INFINITE SEQUENCES AND SERIES, Konrad Knopp. Careful presentation of fundamentals of the theory by one of the finest modern expositors of higher mathematics. Covers functions of real and complex variables, arbitrary and null sequences, convergence and divergence, Cauchy's limit theorem, tests for infinite series, power series, numerical and closed evaluation of series. Translated by Frederick Bagemihl. v + 186pp.   60153-6 Paperbound $2.00

INTRODUCTION TO THE DIFFERENTIAL EQUATIONS OF PHYSICS, Ludwig Hopf. No math background beyond elementary calculus is needed to follow this classroom or self-study introduction to ordinary and partial differential equations. Approach is through classical physics. Translated by Walter Nef. 48 figures. v + 154pp.
60120-X Paperbound $1.75

DIFFERENTIAL EQUATIONS FOR ENGINEERS, Philip Franklin. For engineers, physicists, applied mathematicians. Theory and application: solution of ordinary differential equations and partial derivatives, analytic functions. Fourier series, Abel's theorem, Cauchy Riemann differential equations, etc. Over 400 problems deal with electricity, vibratory systems, heat, radio; solutions. Formerly *Differential Equations for Electrical Engineers*. 41 illustrations. vii + 299pp.
60601-5 Paperbound $2.50

THEORY OF FUNCTIONS, PART II. Single- and multiple-valued functions; full presentation of the most characteristic and important types. Proofs fully worked out. Translated by Frederick Bagemihl. x + 150pp. 60157-9 Paperbound $1.50

PROBLEM BOOK IN THE THEORY OF FUNCTIONS, I. More than 300 elementary problems for independent use or for use with "Theory of Functions, I." 85pp. of detailed solutions. Translated by Lipman Bers. viii + 126pp.
60158-7 Paperbound $1.50

PROBLEM BOOK IN THE THEORY OF FUNCTIONS, II. More than 230 problems in the advanced theory. Designed to be used with "Theory of Functions, II" or with any comparable text. Full solutions. Translated by Frederick Bagemihl. 138pp.
60159-5 Paperbound $1.75

INTRODUCTION TO THE THEORY OF EQUATIONS, Florian Cajori. Classic introduction by leading historian of science covers the fundamental theories as reached by Gauss, Abel, Galois and Kronecker. Basics of equation study are followed by symmetric functions of roots, elimination, homographic and Tschirnhausen transformations, resolvents of Lagrange, cyclic equations, Abelian equations, the work of Galois, the algebraic solution of general equations, and much more. Numerous exercises include answers. ix + 239pp. 62184-7 Paperbound $2.75

LAPLACE TRANSFORMS AND THEIR APPLICATIONS TO DIFFERENTIAL EQUATIONS, N. W. McLachlan. Introduction to modern operational calculus, applying it to ordinary and partial differential equations. Laplace transform, theorems of operational calculus, solution of equations with constant coefficients, evaluation of integrals, derivation of transforms, of various functions, etc. For physics, engineering students. Formerly *Modern Operational Calculus*. xiv + 218pp.
60192-7 Paperbound $2.50

PARTIAL DIFFERENTIAL EQUATIONS OF MATHEMATICAL PHYSICS, Arthur G. Webster. Introduction to basic method and theory of partial differential equations, with full treatment of their applications to virtually every field. Full, clear chapters on Fourier series, integral and elliptic equations, spherical, cylindrical and ellipsoidal harmonics, Cauchy's method, boundary problems, method of Riemann-Volterra, many other basic topics. Edited by Samuel J. Plimpton. 97 figures. vii + 446pp. 60263-X Paperbound $3.00

PRINCIPLES OF STELLAR DYNAMICS, Subrahmanyan Chandrasekhar. Theory of stellar dynamics as a branch of classical dynamics; stellar encounter in terms of 2-body problem, Liouville's theorem and equations of continuity. Also two additional papers. 50 illustrations. x + 313pp. 5⅝ x 8⅜.
60659-7 Paperbound $3.00

CELESTIAL OBJECTS FOR COMMON TELESCOPES, T. W. Webb. The most used book in amateur astronomy: inestimable aid for locating and identifying hundreds of celestial objects. Volume 1 covers operation of telescope, telescope photography, precise information on sun, moon, planets, asteroids, meteor swarms, etc.; Volume 2, stars, constellations, double stars, clusters, variables, nebulae, etc. Nearly 4,000 objects noted. New edition edited, updated by Margaret W. Mayall. 77 illustrations. Total of xxxix + 606pp.
20917-2, 20918-0 Two volumes, Paperbound $5.50

A SHORT HISTORY OF ASTRONOMY, Arthur Berry. Earliest times through the 19th century. Individual chapters on Copernicus, Tycho Brahe, Galileo, Kepler, Newton, etc. Non-technical, but precise, thorough, and as useful to specialist as layman. 104 illustrations, 9 portraits, xxxi + 440 pp.
20210-0 Paperbound $3.00

ORDINARY DIFFERENTIAL EQUATIONS, Edward L. Ince. Explains and analyzes theory of ordinary differential equations in real and complex domains: elementary methods of integration, existence and nature of solutions, continuous transformation groups, linear differential equations, equations of first order, non-linear equations of higher order, oscillation theorems, etc. "Highly recommended," *Electronics Industries*. 18 figures. viii + 558pp.
60349-0 Paperbound $4.00

DICTIONARY OF CONFORMAL REPRESENTATIONS, H. Kober. Laplace's equation in two dimensions for many boundary conditions; scores of geometric forms and transformations for electrical engineers, Joukowski aerofoil for aerodynamists, Schwarz-Christoffel transformations, transcendental functions, etc. Twin diagrams for most transformations. 447 diagrams. xvi + 208pp. 6⅛ x 9¼.
60160-9 Paperbound $2.50

ALMOST PERIODIC FUNCTIONS, A. S. Besicovitch. Thorough summary of Bohr's theory of almost periodic functions citing new shorter proofs, extending the theory, and describing contributions of Wiener, Weyl, de la Vallée, Poussin, Stepanoff, Bochner and the author. xiii + 180pp.
60018-1 Paperbound $2.50

AN INTRODUCTION TO THE STUDY OF STELLAR STRUCTURE, S. Chandrasekhar. A rigorous examination, using both classical and modern mathematical methods, of the relationship between loss of energy, the mass, and the radius of stars in a steady state. 38 figures. 509pp.
60413-6 Paperbound $3.75

INTRODUCTION TO THE THEORY OF GROUP'S OF FINITE ORDER, Robert D. Carmichael. Progresses in easy steps from sets, groups, permutations, isomorphism through the important types of groups. No higher mathematics is necessary. 783 exercises and problems. xiv + 447pp.
60300-8 Paperbound $4.00

ELEMENTARY MATHEMATICS FROM AN ADVANCED STANDPOINT: VOLUME II—GEOMETRY, Feliex Klein. Using analytical formulas, Klein clarifies the precise formulation of geometric facts in chapters on manifolds, geometric and higher point transformations, foundations. "Nothing comparable," *Mathematics Teacher.* Translated by E. R. Hedrick and C. A. Noble. 141 figures. ix + 214pp.

(USO) 60151-X Paperbound $2.25

ENGINEERING MATHEMATICS, Kenneth S. Miller. Most useful mathematical techniques for graduate students in engineering, physics, covering linear differential equations, series, random functions, integrals, Fourier series, Laplace transform, network theory, etc. "Sound and teachable," Science. 89 figures. xii + 417pp. 6 x 8½.

61121-3 Paperbound $3.00

INTRODUCTION TO ASTROPHYSICS: THE STARS, Jean Dufay. Best guide to observational astrophysics in English. Bridges the gap between elementary popularizations and advanced technical monographs. Covers stellar photometry, stellar spectra and classification, Hertzsprung-Russell diagrams, Yerkes 2-dimensional classification, temperatures, diameters, masses and densities, evolution of the stars. Translated by Owen Gingerich. 51 figures, 11 tables. xii + 164pp.

60771-2 Paperbound $2.50

INTRODUCTION TO BESSEL FUNCTIONS, Frank Bowman. Full, clear introduction to properties and applications of Bessel functions. Covers Bessel functions of zero order, of any order; definite integrals; asymptotic expansions; Bessel's solution to Kepler's problem; circular membranes; etc. Math above calculus and fundamentals of differential equations developed within text. 636 problems. 28 figures. x + 135pp.

60462-4 Paperbound $1.75

DIFFERENTIAL AND INTEGRAL CALCULUS, Philip Franklin. A full and basic introduction, textbook for a two- or three-semester course, or self-study. Covers parametric functions, force components in polar coordinates, Duhamel's theorem, methods and applications of integration, infinite series, Taylor's series, vectors and surfaces in space, etc. Exercises follow each chapter with full solutions at back of the book. Index. xi + 679pp.

62520-6 Paperbound $4.00

THE EXACT SCIENCES IN ANTIQUITY, O. Neugebauer. Modern overview chiefly of mathematics and astronomy as developed by the Egyptians and Babylonians. Reveals startling advancement of Babylonian mathematics (tables for numerical computations, quadratic equations with two unknowns, implications that Pythagorean theorem was known 1000 years before Pythagoras), and sophisticated astronomy based on competent mathematics. Also covers transmission of this knowledge to Hellenistic world. 14 plates, 52 figures. xvii + 240pp.

22332-9 Paperbound $2.50

THE THIRTEEN BOOKS OF EUCLID'S ELEMENTS, translated with introduction and commentary by Sir Thomas Heath. Unabridged republication of definitive edition based on the text of Heiberg. Translator's notes discuss textual and linguistic matters, mathematical analysis, 2500 years of critical commentary on the Elements. Do not confuse with abridged school editions. Total of xvii + 1414pp.

60088-2, 60089-0, 60090-4 Three volumes, Paperbound $9.50

ASTRONOMY AND COSMOGONY, Sir James Jeans. Modern classic of exposition, Jean's latest work. Descriptive astronomy, atrophysics, stellar dynamics, cosmology, presented on intermediate level. 16 illustrations. Preface by Lloyd Motz. xv + 428pp. 60923-5 Paperbound $3.50

EXPERIMENTAL SPECTROSCOPY, Ralph A. Sawyer. Discussion of techniques and principles of prism and grating spectrographs used in research. Full treatment of apparatus, construction, mounting, photographic process, spectrochemical analysis, theory. Mathematics kept to a minimum. Revised (1961) edition. 110 illustrations. x + 358pp. 61045-4 Paperbound $3.50

THEORY OF FLIGHT, Richard von Mises. Introduction to fluid dynamics, explaining fully the physical phenomena and mathematical concepts of aeronautical engineering, general theory of stability, dynamics of incompressible fluids and wing theory. Still widely recommended for clarity, though limited to situations in which air compressibility effects are unimportant. New introduction by K. H. Hohenemser. 408 figures. xvi + 629pp. 60541-8 Paperbound $5.00

AIRPLANE STRUCTURAL ANALYSIS AND DESIGN, Ernest E. Sechler and Louis G. Dunn. Valuable source work to the aircraft and missile designer: applied and design loads, stress-strain, frame analysis, plates under normal pressure, engine mounts, landing gears, etc. 47 problems. 256 figures. xi + 420pp. 61043-8 Paperbound $3.50

PHOTOELASTICITY: PRINCIPLES AND METHODS, H. T. Jessop and F. C. Harris. An introduction to general and modern developments in 2- and 3-dimensional stress analysis techniques. More advanced mathematical treatment given in appendices. 164 figures. viii + 184pp. 6⅛ x 9¼. (USO) 60720-8 Paperbound $2.50

THE MEASUREMENT OF POWER SPECTRA FROM THE POINT OF VIEW OF COMMUNICATIONS ENGINEERING, Ralph B. Blackman and John W. Tukey. Techniques for measuring the power spectrum using elementary transmission theory and theory of statistical estimation. Methods of acquiring sound data, procedures for reducing data to meaningful estimates, ways of interpreting estimates. 36 figures and tables. Index. x + 190pp. 60507-8 Paperbound $2.50

GASEOUS CONDUCTORS: THEORY AND ENGINEERING APPLICATIONS, James D. Cobine. An indispensable reference for radio engineers, physicists and lighting engineers. Physical backgrounds, theory of space charges, applications in circuit interrupters, rectifiers, oscillographs, etc. 83 problems. Over 600 figures. xx + 606pp. 60442-X Paperbound $3.75

Prices subject to change without notice.

Available at your book dealer or write for free catalogue to Dept. Sci, Dover Publications, Inc., 180 Varick St., N.Y., N.Y. 10014. Dover publishes more than 150 books each year on science, elementary and advanced mathematics, biology, music, art, literary history, social sciences and other areas.